Metallurgy in Ancient Ecuador

A Study of the Collection of Archaeological Metallurgy of the Ministry of Culture, Ecuador

Roberto Lleras Pérez

Archaeopress Precolumbian Archaeology 5

ARCHAEOPRESS PUBLISHING LTD
SUMMERTOWN PAVILION
18-24 MIDDLE WAY
OXFORD OX2 7LG

www.archaeopress.com

ISBN 978 1 78491 160 7
ISBN 978 1 78491 161 4 (e-Pdf)

© Archaeopress and R Lleras Pérez 2015

All rights reserved. No part of this book may be reproduced, stored in retrieval system, or transmitted, in any form or by any means, electronic, mechanical, photocopying or otherwise, without the prior written permission of the copyright owners.

This book is available direct from Archaeopress or from our website www.archaeopress.com

Contents

Contents i

Resumen en español v

Acknowledgements vii

Introduction 1

The collection of the Ministry of Culture 5

Previous studies on the pre-Hispanic metallurgy of Ecuador 10

Metallogenesis and metal resources in Ecuador 21

Early finds and the Initial Period 28

Great Regional Groups: La Tolita –Tumaco 32
    Geographic Distribution 32
    Chronology 35
    Technology 36
    Typology and classification 42

Great Regional Groups: Jama – Coaque 61
    Geographical distribution 61
    Chronology 63
    Technology 63
    Typology and classification 64

Great Regional Groups: Bahia 70
    Geographic distribution 70
    Chronology 72
    Technology 73

Great Regional Groups: Milagro – Quevedo 81
    Geographic Distribution 81
    Chronology 82
    Technology 84
    Typology and classification 88

Great Regional Groups: Manteño - Huancavilca ... 97
- Geographic Distribution ... 97
- Chronology ... 99
- Technology ... 100
- Typology and classification ... 101

Great Regional Groups: Puruha ... 111
- Geographic Distribution ... 111
- Chronology ... 113
- Technology ... 113
- Typology and classification ... 116

Great Regional Groups: Cañari ... 127
- Geographic Distribution ... 127
- Chronology ... 129
- Technology ... 129
- Typology and classification ... 131

Great Regional Groups: Carchi – Nariño ... 138
- Geographic Distribution ... 138
- Chronology ... 142
- Technology ... 143
- Typology and classification ... 146

Isolated finds and problematic Groups ... 167
- The Coast ... 167
- The Sierra ... 170
- Discussion ... 171

The Inca metallurgical integration ... 174
- Geographic Distribution ... 175
- Technology ... 178
- Typology and classification ... 179

Iconography and symbolism in metallurgy ... 188

Synthesis ... 195

An interpretative proposal for the development of metallurgy in Ecuador ... 200

References ... 203

List of Figures

Figure 1 Museo Nacional del Ecuador in Quito, house of the collection of pre-Hispanic archaeological metal objects.. 8
Figure 2 Provenances of metallic objects of the collection of the Ministry of Culture of Ecuador 9
Figure 3 Olaf Holm, one of the pioneers of the study of metal artefacts in Ecuador 20
Figure 4 Chimborazo the highest strato-volcano in Ecuador; metal deposits are associated to volcanic activity. 26
Figure 5 Alluvial river placers like this one in the lowlands of the Pacific coast abound in gold and platinum. 27
Figure 6 Provenances of La Tolita - Tumaco... 33
Figure 7   Provenance of La Tolita - Tumaco metal objects in southern Colombia 34
Figure 8 La Tolita – Tumaco gold anthropomorphic mask with extensions imitating the rays of the sun... 51
Figure 9 La Tolita – Tumaco gold and platinum zoomorphic mask. 52
Figure 10 La Tolita – Tumaco gold and platinum with sodalite inlays anthropomorphic mask 53
Figures 11 and 12   La Tolita – Tumaco gold ear pendants 54
Figure 13 La Tolita – Tumaco gold and platinum zoomorphic mask, two components 55
Figure 14 La Tolita – Tumaco gold and platinum anthropomorphic mask................................. 56
Figure 15 La Tolita – Tumaco gold crest for diadem. 57
Figure 16 La Tolita – Tumaco gold necklace.. 58
Figure 17 La Tolita – Tumaco gold zoomorphic figure. 59
Figure 18 La Tolita – Tumaco copper axe 60
Figure 19 Provenances of Jama - Coaque metal objects 62
Figure 20 Jama – Coaque gold pendants.. 66
Figure 21 Jama – Coaque gold bowl 67
Figures 22 and 23 Jama – Coaque gold ear pendants 68
Figure 24 Jama – Coaque gold breastplate with zoomorphic figure 69
Figure 25 Provenance of Bahia metal objects 71
Figure 26 Bahia silver votive figure shaped as a raft 77
Figure 27 Bahia gold snail cover 78
Figure 28 Bahia silver chest guard 79
Figure 29 Bahia gold pair of ear pendants.. 80
Figure 30 Provenance of Milagro - Manatňo 83
Figure 31 Milagro – Quevedo copper crucible 91
Figure 32 Milagro – Quevedo copper mould 92
Figure 33 Milagro – Quevedo copper staff: 93
Figure 34 Milagro – Quevedo copper axe-monies 94
Figure 35 Milagro – Quevedo gold nose ornament. 95
Figure 36 Milagro – Quevedo gold spiral nose ornament. 96
Figure 37 Provenances for Manteño – Huancavilca metal objects... 98
Figure 38 Manteño – Huancavilca silver and copper mask with crown.................................... 105
Figure 39 Manteño – Huancavilca silver and copper mask with crown 106
Figure 40 Manteño – Huancavilca silver breastplate 107
Figure 41 Manteño – Huancavilca silver plaque 108
Figure 42 Manteño – Huancavilca copper axe.. 109
Figure 43 Manteño – Huancavilca copper breastplate, tinculpa style 110
Figure 44 Provenance of Puruha metal objects... 112
Figure 45 – Puruha gold spear throwers 120
Figure 46 – Puruha giant copper *tupo*.. 121
Figure 47 – Puruha copper crown 122
Figures 48 and 49 – Puruha gold ear pendants with zoomorphic figures. 123
Figures 50 and 51 – Puruha gold and silver ear pendants 124

Figure 52 Puruha gold and silver nose ornament. 125
Figure 53 Puruha gold anthropomorphic pendant. 126
Figure 54 Povenance of Cañari metal objects 128
Figure 55 Cañari gold diadem. 134
Figure 56 Cañari gold with Spondylus inlays ear pendant lid 135
Figure 57 Cañari gold ear pendant lid. 136
Figure 58 Cañari gold pendant with anthropomorphic figures. 137
Figure 59 Provenenace of Carchi – Nariño metal objects 139
Figure 60 Figure 59 Provenenace of Carchi – Nariño metal objects in southern Colombia 141
Figure 61 Carchi - Nariño gold mask 159
Figure 62 Carchi - Nariño tombac ear pendants, tinculpa style: 160
Figure 63 Carchi - Nariño gold ear pendants with zoomorphic figures 161
Figure 64 Carchi - Nariño tombac nose ornament. 162
Figure 65 Carchi - Nariño tombac necklace. 163
Figure 66 Carchi - Nariño gold pendant shaped as a bird 164
Figure 67 Carchi - Nariño gold pendant shaped as a bird 165
Figure 68 Carchi - Nariño tombac breastplate 166
Figure 69 Provenance of metal objects with no cultural attribution 168
Figure 70 Provenance of Inca metal objects 177
Figure 71 Inca silver ceremonial vase (*kero*) 182
Figure 72 Inca silver arivaloid bottle. 183
Figure 73 Inca gold and silver votive figures. 184
Figure 74 Inca gold *tupo* 185
Figure 75 Inca bronze axe 186
Figure 76 Inca bronze head breaker 187

All the photographs are the property of the *Ministerio de Cultura y Patrimonio del Ecuador* and *Archivo Histórico del Ministerio de Cultura y Patrimonio del Ecuador* that have granted permission for their use in this book. Editor of the *Fondo Editorial*: Adriana Grijalva. Photographers: Jorge Delgado, Patricio Estévez and Santiago Ontaneda.

Resumen en español

Este trabajo surgió como respuesta a la necesidad de renovar la Sala de Metales Arqueológicos del Museo Nacional del Ecuador. Para ello se recopilaron y sistematizaron los conocimientos sobre metalurgia precolombina del Ecuador. Paralelamente se estudió la colección de metalurgia del Banco Central (hoy Ministerio de Cultura) de Quito. Este estudio comprendió: a) El examen de todas las piezas; b) La inclusión de estos objetos en una base de datos; c) La sistematización de las informaciones de procedencia y su manejo en mapas georreferenciados; d) La recopilación de los fechados radiocarbónicos asociados con metalurgia y e) La realización de análisis metalúrgicos y metalográficos de muestras seleccionadas.

Los autores anteriores definieron culturas, estilos o tradiciones. En este trabajo se adoptó la clasificación usada por el Museo Nacional que comprende las siguientes categorías: a) La Tolita; b) Jama-Coaque; c) Bahía; d) Milagro-Quevedo; e) Manteño-Huancavilca; f) Puruhá; g) Cañari; h) Negativo del Carchi e i) Inca. Sobre este sistema se realizaron las siguientes modificaciones: a) Se adoptó para la Costa Norte el apelativo La Tolita-Tumaco para incluir la costa del Pacifico sur de Colombia, que fue una misma área cultural con la costa norte del Ecuador, así se incluyeron las piezas del Museo del Oro de Colombia de esta procedencia; b) Lo mismo se hizo para la Sierra Norte que se denominó Carchi-Nariño, ingresando las piezas del departamento de Nariño y sur del departamento del Cauca en Colombia; c) Dada la incertidumbre sobre si las categorías clasificatorias conforman culturas, etnias, estilos, tradiciones, periodos o fases se optó por el uso de una categoría más neutra. Esta categoría se denominó Gran Conjunto Regional y cada uno de ellos fue caracterizado en términos de: a) Su área geográfica de dispersión; b) La definición de su cronología; c) La tecnología; d) La tipología de forma función; e) La iconografía y simbología.

Por tratarse de un estudio basado en objetos de colección, en su mayor parte descontextualizados, hay limitaciones. La información de procedencia está afectada por factores como la imprecisión en la ubicación de los lugares de hallazgo, la omisión de esta información y las mezclas de piezas de diversa procedencia en un mismo lote. La ausencia de contextos arqueológicos limitó nuestra comprensión de la asociación interna de piezas dentro de los ajuares, así como la posibilidad de comparar entre si conjuntos de diversas tumbas de un mismo sitio. La información arqueológica, contenida en la bibliografía, también tiene sus limitaciones. Hay diferencias entre distintas áreas geográficas en cuanto a la cantidad y calidad de las investigaciones arqueológicas que han tratado la metalurgia. Lo que resultó muy productivo fue la combinación de la información de la arqueología de campo con la del estudio de colecciones. Este libro hace un balance actualizado de los conocimientos arqueológicos, tecnológicos, sociopolíticos, estéticos y simbólicos sobre la industria metalúrgica precolombina en el actual territorio del Ecuador y propone una interpretación global de su simbología e iconografía, así como su probable desarrollo histórico desde sus orígenes hasta la conquista europea en el siglo XVI.

Acknowledgements

To be able to do this research, to enter the fascinating world of the pre-Columbian metallurgy of Ecuador and to have the privilege of examining closely thousands of marvellous metal objects have been some of the most grateful experiences of my professional life. I must recognise that this was possible because Santiago Ontaneda stubbornly insisted that I should write the layout for the exhibition of the *Sala de Metales Arqueologicos* (Room of Archaeological Metals) in the *Museo Nacional del Banco Central*, in spite of several obstacles and problems that arose at the time; for this and for his support and friendship I am deeply indebted.

Equally important was the open confidence, offered both by Fernando Moncayo and Carlos Landazuri, Directors of Culture of the *Banco Central*, who issued the administrative decision that enabled my participation and honoured me with their kindness. Without the enthusiasm, generous help, knowledge and advice of Estelina Quinatoa the examination of the objects in the deposits would not have been so successful. Within the team of the Cultural Direction of the Banco Central I must also recognise the help of Roberto Cárdenas, Johnny Hidalgo, Gustavo Espíndola, Antonio Fresco and Marcelo Calderón. Patricia Estévez, who treasures in her heart the Indian metallurgy of Ecuador, helped me with information and concepts that are to be found only in her perceptive mind. Luís Oquendo, another lover of pre-Hispanic metals, shared with me his deep knowledge of basic materials. Mercedes Guinea, Francisco Valdez and Maria del Carmen Molestina gave me access to unpublished articles and data.

To get hold of the bibliography composed of over a hundred references, some of them dating from the 19th century, was a difficult task in which I had much help from Sandra Mendoza, Laura Lleras, Jorge Morales, Juanita Sáenz and Luz Alba Gómez. Luz Alba also allowed me to use her excellent works on the archaeology of Nariño, Colombia. Assembling the initial lists of registration numbers for the database was a long and tedious work that Clara Sterling undertook with pleasure and dedication. Jaime Arias gave me valuable advice for the edition of the geo-referenced maps.

The *Departamento Técnico Industrial* (Technical Industrial Department) of the *Banco de la Republica* of Colombia carried out the analysis of samples whose composition and characteristics we needed to know in order to solve specific problems of the study; those analyses were done by Carlos Hernández, Nohra Bustamante, Lisette Garzón and Hernán Arias. Thanks to the support of the *Fundación de Investigaciones Arqueológicas Nacionales of Colombia* it was possible to obtain several radiocarbon dates. I had valuable help from the *Instituto Geográfico Militar del Ecuador* to locate some archaeological sites.

The research work would not have been possible without the support of the directors of the *Banco de la República* who understood the need for this collaborative work and the mutual benefits that would derive from it. During the final phase of the preparation of the manuscript I received valuable advice from Adriana Grijalva. I must also thank the superb photographic work of Jorge Delgado.

Last but not least; it was Paul Bahn who undertook the job of finding a publisher in the United Kingdom. By doing this he closed the circuit that gave birth to this book, thank you very much!

Bogota, June 2015

Introduction

The research work published in this book started when it became evident that the exhibition and the scientific layout of the *Sala de Metales Arqueologicos* (Room of Archaeological Metals) of the *Museo Nacional* of the *Banco Central del Ecuador* (nowadays *Ministerio de Cultura* (Ministry of Culture)) were in urgent need of renewal. The previous layout, still mounted when this work started in 2005, dated from 1995 and had been changed only slightly; it assembled the research results obtained previously with the classification, museographic layout and scientific analyses of the collection of the *Banco Central*. The archaeological background was based on the general scheme of Ecuadorian archaeology as it is explained, for example, in the classic text "Ecuador" (Meggers, 1966). This synthesis gathers the results and hypothesis of many previous researchers and has been adopted since then with few variations by most specialists of the region.

In order to achieve the goal of renewing the scientific layout it was considered of fundamental importance to collect and systematise the existing general knowledge about pre-Hispanic metallurgy of Ecuador and the specific data concerning the collection of the *Banco Central*; something that had not been done before. The first stage implied accessing and revising over a hundred related bibliographical references. These references are to be found in books, articles, catalogues, manuscripts, excavation reports, laboratory reports, projects, exhibition layouts and travel chronicles. Only a fraction of these documents is entirely dedicated to the pre-Hispanic metallurgy of Ecuador; even though most of them have only isolated remarks or theoretical considerations about metallurgy they are extremely valuable. Specialists in Ecuadorian archaeology might notice that certain books that we decided not to include in this revision might contain marginal references to metallurgy; though this is right, it is true also that these exclusions do not affect seriously the general panorama and that most of those books are cited in one or other of the texts consulted.

Concurrently we carried out the study of the metallurgy collection in the deposits of the *Banco Central* and in the exhibition of the Room of Archaeological Metals of the *Museo Nacional* in Quito. This study was structured according to the following methodology: a) Visual examination, sometimes restricted, of all the available objects; b) Inclusion of those objects in a database that included as much information as could be gathered for each record; c) Organisation of the provenance information in geo-referenced maps so as to allow us to define and clearly show the distribution patterns; d) Compilation and organisation of radiocarbon dates associated to metallurgy, either obtained from collection objects or in archaeological digs and e) Processing of metallurgical and metallographic analyses of samples selected according to their importance or the absence of specific knowledge.

When dealing with a field of study that has not been subjected previously to an integrated analysis we cannot take anything for granted. In the case of Ecuadorian pre-Hispanic metallurgy, despite the fact that many topics had been already examined, the basic axes of archaeological work were still ill-defined. We thought that this research should aim at solving four basic questions: 1) Where? That meant defining as precisely as possible the spatial axis of metallurgy; 2) When? That meant establishing the general and the regional chronologies; 3) How? That meant describing the technology of manufacture; 4) What? That meant defining the formal and functional repertoire and the iconography. Obviously there are many other very important aspects that are not included in these aims, but it will be easier now to approach those topics with a solid backup of organised archaeo-metallurgical information.

Those authors that, at different moments, have published on the topic of pre-Hispanic metallurgy of Ecuador have proposed the existence of diverse cultures, styles or traditions. Due to the nature of our work we decided to adopt the classificatory system used by the *Museo Nacional* in its archaeological deposits and in the Gold Room of the museum. It is important to stress that no major changes were introduced in this system because it was not considered feasible to do so at the time. This classification system includes the following major cultures: 1) La Tolita; 2) Jama-Coaque; 3) Bahia; 4) Milagro-Quevedo; 5) Manteño-Huancavilca; 6) Puruha; 7) Cañari; 8) Negativo del Carchi and 9) Inca. Some cultures that appear occasionally but are not systematically and coherently used were assimilated to the major ones; that is the case with Guangala, Tacalshapa and Narrio.

The classificatory system was supplemented with the following additions that do not alter its basic principles: a) We changed the term used for the northern coast to La Tolita - Tumaco so as to include in the distribution area the southern Pacific coast of Colombia, taking into account that in pre-Hispanic times both were just one cultural area. Then we included in the database the objects from the collection of the *Museo del Oro* found in the south Pacific coast of Colombia; b) A similar operation, based on the same considerations, was performed for the northern Sierra that was then re-named as Carchi-Nariño. In the database we then included the objects found in the Andean region of the department of Nariño and the southern portion of the Cauca department in Colombia; c) In order to overcome the uncertainty caused by the fact that it was very difficult to define whether the classificatory categories (cultures) really corresponded to cultures or to ethnic groups, styles, traditions, periods or phases or if those criteria were intermingled, we adopted a neutral category that does not force the study within any particular theoretical approach and leaves the door open for alternative interpretations. In this study the classificatory categories are termed as Great Regional Groups and they do not imply any notion of ethnic identity; nevertheless when a safe correlation can be established between a known Indian group and a metallurgical group, that link is clearly expressed.

Each Great Regional Group was defined in terms of its constitutive characteristics, those that make it unique and different from the other groups. The set of characteristics include: a) The definition of its geographical area of distribution, that is the region where the objects were made and used; b) The definition of its chronology, the period during which the tradition was in vigour; c) The technology, the repertoire of manufacturing and finishing techniques used to make the objects; d) The form and function typology, that is the range of types and shapes of objects produced; e) The iconography and symbology, individual decorative motifs and their combinations used to articulate a particular symbolic system.

This characterisation does not exhaust the field of study. It would be important to go more in depth into each of the aspects considered here and it would also be vital to take into account other topics that we were not able to include at this stage, such as the definition of the relations between the regional groups of metallurgy and the pottery types and styles as well as the compilation of information concerning the archaeological contexts of metal finds.

Our formal, functional, technological and iconographic classification followed closely the criteria established by the *Banco Central*, especially the inventory made by researcher Patricia Estévez that contains almost all the objects of the collection. This inventory, together with the visual examination and the photographic record allowed feeding the database. It is important to note that the shortage of time prevented the direct comparison of each and every object against the records.

Due to the fact that this study dealt with objects belonging to a museum collection, most of them lacking context, there are restrictions which must be acknowledged. The provenance information, indispensable to draw distribution areas and to understand the extension and magnitude of exchanges, is seriously affected by facts such as the imprecise location of find sites, the absence of such information, the mixing up of objects from different provenances in one acquisition batch reported as coming from only one site, etc. This means that the distribution zones should be seen as indicative but not as sharply defined territories.

From another point of view the absence of archaeological contexts severely limited our understanding of the internal association of objects within funerary assemblages and ritual attires, as well as the possibility of comparing sets coming from different graves within the same site. Those and several other problems are usually found by archaeologists studying museum collections. There are, however, advantages that offset these problems and render viable and productive the study of collections without contexts. The mere accumulation of objects is, per se, a positive factor. The examination of large quantities of objects is the nearest way to reach the ideal of a complete vision of the technological, formal, functional decorative, iconographic and symbolic repertoire.

The study of technology is not generally affected by the fact that objects belong to collections. It is important to say that we should not avoid the responsibility of studying collections just because they do not offer the ideal conditions that the archaeologist is trained to recover in the field.

The archaeological information contained in the revised references is also restricted. There are noticeable differences between the regions with respect to the quantity and quality of the archaeological projects that have dealt with metallurgy. Whereas the La Tolita-Tumaco area has been visited by many research expeditions, the region of Chimborazo (Puruha) has been left mostly untouched. The authors do not always give the metal objects found in the course of their digs all the attention they deserve, mainly because they are not the objective of their investigations. Ultimately what proved to be most productive was to combine the information coming from field archaeology with the study of collections. However, not even this combination managed to answer all of our queries, but it does improve the general perspective.

This balance of our present knowledge of the archaeological, technological, socio-political, aesthetic and symbolic aspects of pre-Hispanic metallurgy of Ecuador is aimed also at establishing the origins of this industry and its relations with neighbouring areas, especially southern Colombia. We also intended to throw some light upon the survival of metal working after the Spanish conquest.

The database "Pre-Hispanic Metallurgy of Ecuador" (Microsoft Access software) is an integral part of this document. This database has a total of 7786 records of objects in a main table with 5179 records of objects from the *Banco Central del Ecuador* (Quito) and a secondary table with 2607 objects from southern Colombia presently in the *Museo del Oro,* Colombia. Another component of the document is a series of nine geo-referenced maps (an IGME document modified with ESRI ArcGis software) that show the general distribution of provenances and the particular distribution of each Great Regional Group.

Chapter 1

The collection of the Ministry of Culture

The *Banco Central del Ecuador*, as many other central banks in the Andean countries where precious metals are abundant, undertook right from its beginning the acquisition, refinement and custody of national gold. This was the reason why many archaeological objects mixed by miners or treasure hunters with alluvial or mine gold found their way into the laboratories and offices of the bank (Zapater, n.d). Prior to 1950 and thanks to the officer in charge of the laboratory, Dr. Arauz, archaeological metal objects were separated from the raw metal, thus avoiding their melting down in the furnaces; with them began a small collection that was to be the core of the future museum (Zapater, n.d).

In the decade of 1950 the bank undertook the preservation of archaeological objects as an official policy and assigned resources destined to buy objects excavated by treasure hunters or belonging to private collectors. The great quantitative and qualitative leap forward occurred in 1960 when the bank's authorities decided to buy what was then the most important private collection in Ecuador (Zapater, nd). The objects bought from Max Konanz, over a thousand counting metals only, transformed the collection of the *Banco Central* in an invaluable heritage with a huge potential for research and exhibition. The possibility of building a museum representing the pre-Hispanic cultures of Ecuador thus arose and became stronger progressively.

From then onwards and during the years from 1960 to 2005 the collection kept on growing steadily with variable annual frequencies that seem to reflect the equally variable interest of the bank's authorities for that type of objects. At present the museum and deposit collection in Quito, now under the administration of the Ministry of Culture, has 5179 objects that have been acquired as shown:

Year	Objects acquired in each year	Annual acquisitions as percentage of total	Accumulated growth of the collection	Accumulation as percentage of the total
1946	1	0.02	1	0.02
1947	26	0.50	27	0.52
1948	32	0.62	59	1.14
1949	59	1.14	118	2.28
1950	6	0.12	124	2.40
1957	12	0.23	136	2.63
1959	240	4.63	376	7.26
1960	1141	22.04	1517	29.30

Year	Objects acquired in each year	Annual acquisitions as percentage of total	Accumulated growth of the collection	Accumulation as percentage of the total
1962	70	1.35	1587	30.65
1963	142	2.74	1729	33.39
1964	111	2.14	1840	35.53
1965	122	2.36	1962	37.89
1966	196	3.79	2158	41.68
1967	14	0.27	2172	41.95
1968	98	1.89	2270	43.84
1969	353	6.82	2623	50.66
1970	641	12.38	3264	63.04
1971	324	6.26	3588	69.30
1972	227	4.39	3815	73.69
1973	108	2.09	3923	75.78
1974	96	1.85	4019	77.63
1975	72	1.39	4091	79.02
1976	141	2.72	4232	81.74
1977	235	4.54	4467	86.28
1978	113	2.18	4580	88.46
1979	60	1.16	4640	89.62
1980	98	1.89	4738	91.51
1981	69	1.33	4807	92.84
1982	53	1.02	4860	93.86
1983	6	0.12	4866	93.98
1984	46	0.89	4912	94.87
1985	10	0.19	4922	95.06
1988	22	0.42	4944	95.48
1992	13	0.25	4957	95.73
1995	4	0.08	4961	95.81
1996	6	0.12	4967	95.93
1997	21	0.41	4988	96.34
1998	1	0.02	4989	96.36
1999	8	0.15	4997	96.51
2001	55	1.04	5052	97.55
2002	10	0.19	5062	97.74
2003	19	0.37	5081	98.11
2005	98	1.89	5179	100

In 1969 the new venue of the museum was opened in one of the buildings of the *Banco Central* in Quito. What was previously a precarious and limited exhibition became, thanks to a research and museographic effort, a modern cultural centre that no longer showed just objects but rather a scientific and aesthetic discourse about the past of Ecuador (Zapater, nd.).

A few years after, in 1973, international travelling exhibitions were being sent to various countries. There followed the strengthening of the educational programmes and the archaeological research and the de-centralisation of the exhibition, a task that led to the opening of the regional museums and site museums. In 2010 the Central Bank (now Ministry of Culture) has archaeological exhibitions that include metal artefacts in ten venues located in the cities of Quito, Guayaquil, Cuenca, Loja, Riobamba, Esmeraldas, Bahía, Manta, Ibarra and the site museum of Ingapirca. Those exhibitions bring archaeological themes to local and regional populations thus accomplishing a vital educational function.

The present *Museo Nacional*, setting of the most important archaeological exhibition in Ecuador, was opened to the public in 1995 (Zapater, nd). The scientific layout is, as previously stated, a complete and comprehensive synthesis of the country´s archaeology that served as the basis for this work.

The effort displayed by the institution since the decision was taken to acquire, preserve and exhibit archaeological metallic objects is extremely relevant. The following table summarises the quantity of objects that has been saved from destruction or illegal export and that are part of the collection of the *Banco Central* (Ministry of Culture) in the city of Quito.

Regional Group	Number of objects	Percentage of total
La Tolita-Tumaco	1,741	33.62
Jama-Coaque	174	3.36
Bahía	295	5.70
Manteño-Huancavilca	340	6.57
Milagro-Quevedo	372	7.18
Carchi-Nariño	479	9.25
Puruhá	258	4.96
Cañari	112	2.16
Inca	180	3.48
Hispanic-Indian	5	0.10
Undetermined attribution	1,223	23.62
Total	5,179	100

Figure 1 Museo Nacional del Ecuador in Quito, house of the collection of pre-Hispanic archaeological metal objects

The collection of the Ministry of Culture

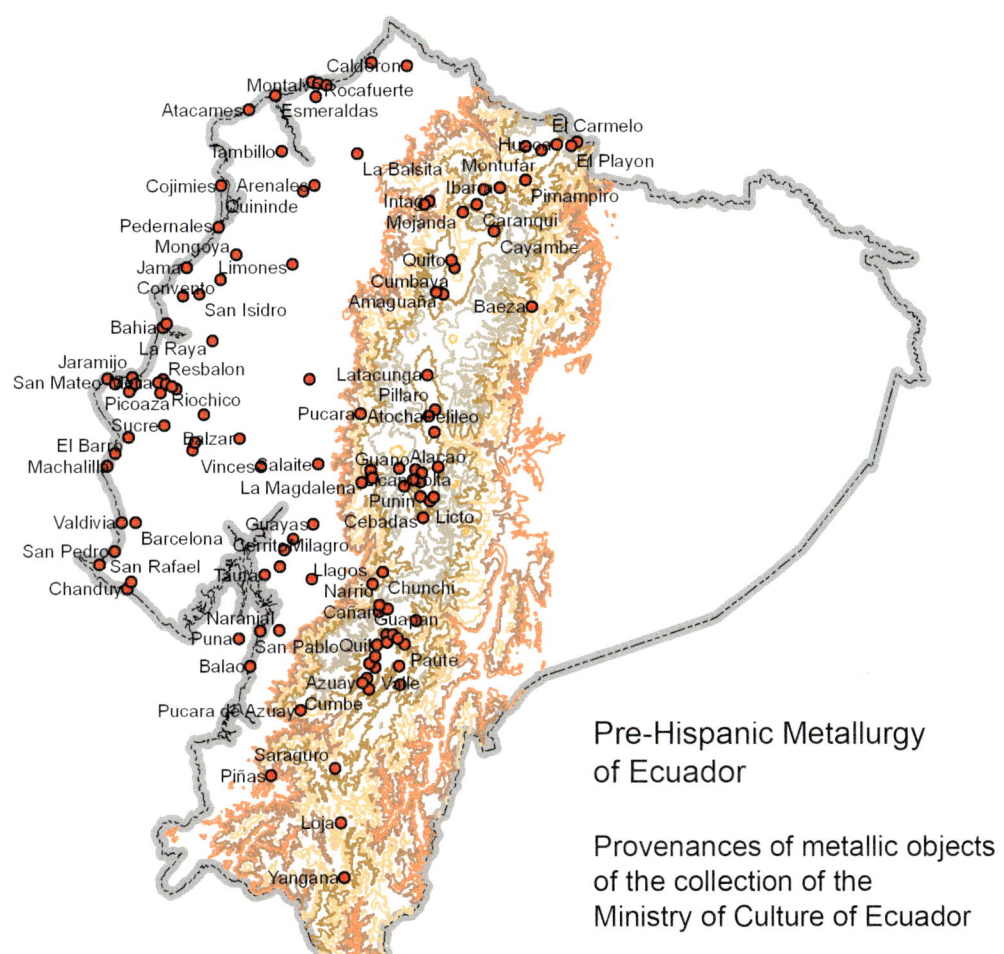

Pre-Hispanic Metallurgy of Ecuador

Provenances of metallic objects of the collection of the Ministry of Culture of Ecuador

Chapter 2

Previous studies on the pre-Hispanic metallurgy of Ecuador

The metallurgy of ancient Ecuador was appealing to the conquistadors from the moment when the first contact between them and the Indians occurred; the famous episode of the crossing between the Spanish ships of Bartolomé Ruiz and the raft of the Huancavilca merchants opened the door to the knowledge of this industry and also to the plundering of the gold of the Indians (Holm 1970). Almost immediately the pillage of Coaque took place (1531) followed by the dispatch of the stolen gold to Spain (Holm 1970). During the conquest of the Ecuadorian coast these actions were frequently repeated, while in the Sierra the dismembering of the Inca Empire and the dismounting of its administrative structure came along with the expropriation of the precious metals owned by the Indians. Among the most famous episodes we can cite the plundering of a rich tomb located near Ingapirca, which was perpetrated by a group of Spaniards in 1563, producing booty comparable to the annual produce of one the largest *encomiendas* of the region (Salomón 1987 in Hosler et al 1990).

Once the Spanish colonial authority was well established the hunger for gold of the new masters had to be satisfied by means of the exploitation of more stable sources; mining carried out with the forced labour of the Indians managed to extract from the mountains of the Sierra (Rodicio et al 1998) and the gold sands of the coastal rivers large quantities of metal that, together with the booty from Peru, New Spain and New Granada, fed the Royal Treasure of Spain for decades. During this long period, over three centuries, the common denominator was the absolute disdain for the aesthetic, technical and symbolic qualities of the metal objects of the Indians. Not one of the extraordinary finds of that time was preserved and there are no acceptable descriptions of them; with no exception they all went into the melting furnace. There are few references to the god of the Indians, such as the one of Lope de Atienza who wrote that the small axes in Cañari served as exchange goods in the marriages (in Hosler et al 1990). The only text written at the time on metallurgy is to be found in the work of Benzoni *"La Historia del Mondo Nuevo"* (1565/1985) which, in just a paragraph, roughly described how the Quito smiths made their gold and silver objects.

Even though the fate of several well-known finds of the 19th century was similar, fortunately the spirit of the time was quite another thing. Thanks to books such as those by Heuzey: *"Le Tresor de Cuenca"* (1870); Bastian *"Die Culturländer des Alten America"* (1878-86); Bamps *"Les antiquites equatoriennes du Musee Royal d'Antiquites de Bruxelles"* (1879); Wolf *"Viajes Científicos por la Republica del Ecuador, III.*

Memorias sobre geografía y geología de la provincia de Esmeraldas" (1879) and González Suárez *"Prehistoria Ecuatoriana"* (1904) we know of tombs found in El Angel, Cuenca, Patecte, Sigsig, Guapán and other sites that had very important assemblages of metal objects (Mayer 1992) and about the objects that were taken to European museums that were examined with certain attention. The descriptions and drawings left at that time, despite their lack of accuracy and thoroughness, show the awakening of a remarkable interest in the archaeological metals and prove that they were considered as a valid field of study. That this was true us attested by the fact that a collection of Ecuadorian archaeological objects was sent to Chile to be shown in the International Exhibition of 1875. The catalogue, wrote by Rencoret, includes the description of an extraordinarily rich tomb found in Guapán (Hosler et al 1990)

Most of the people who first set to find and collect systematically metallic objects were foreigners. During the first decade of the 20th century several missions sent by North American and European museums, as well as adventurers that came by themselves, explored and exploited important sites in the Coast and Sierra. This is the case of Dorsey *"Archaeological investigations on the island of La Plata, Ecuador"* (1901); Saville *"The Antiquities of Manabí, Ecuador. Contributions to South American Archaeology"* (1907-10) and *"The Gold Treasure of Sigsig, Ecuador"* (1924), mentioned by Segarra in *"Comentarios y acotaciones al tesoro de Sigsig, Ecuador"*; Von Buchwald *"Notas acerca de la arqueología del Guayas"* (1918) and Farabee *"A Golden Hoard from Ecuador"* (1921). Most of the objects found at the time are kept in the museums that sponsored the expeditions; some went into private collections and, finally, a few after much wandering, ended in the national museums. The descriptions of tombs, the layout of attires and archaeological contexts lack detail and precision but are, nonetheless, valuable and should be consulted when studying pre-Hispanic metallurgy.

The arrival in Ecuador of some important pioneers of archaeology and anthropology kindled a substantial change in the course of the studies. With Uhle, Verneau, Arsandaux and Rivet genuine research expeditions begin to accumulate contextual information on metallurgy. There are also enquiries into technology, a hitherto ignored subject. Max Uhle in *"Sepulturas ricas de oro en la Provincia del Azuay"* (1922) described the contents of the famous tombs of Chordeleg, Sigsig and other sites; Verneau and Rivet in *"Ethnographie ancienne de l'Equateur"* (1912) recorded and analysed metallic objects from several provenances; Rivet and Arsandaux in *"La Metallurgie en Amerique Précolombienne"* (1946) characterised the metallic traditions of Ecuador in relation to the neighbouring regions. Reichlen in *"Contribution a l'etude de la métallurgie Précolombienne de la province d'Esmeraldas (Equateur)"* (1942) proposed a typology based on collections and included a few preliminary metallurgical analyses.

At approximately the same time the great pioneer of Ecuadorian archaeology, Jacinto Jijon y Caamaño, was advancing on his general studies of Ecuador that involved theoretical considerations and descriptions of metal finds, as in *"Antropología Prehispánica del Ecuador"* (1997). He also carried out specific studies on metallurgy as the one devoted to the group of pendants known as tinculpas: *"Las tinculpas y notas acerca de la metalurgia de los aborígenes del Ecuador"* (1920); Jijon y Caamaño also addressed the problem of bronze in America in his *"La Edad de Bronce en América del Sur"* (1922). In the meanwhile historians as Miguel Duran continued recording famous finds: *"Entierros en Huapan"* (1930).

During the 1930s and 1940s this thrust becomes stronger; that is when Paul Bergsoe publishes two studies that marked a key point in the study of the native metallurgical technology of platinum and copper: *"Metalurgia y tecnología de oro y platino entre los indios precolombinos"* (1937) and *"The gilding process and the metallurgy of copper and lead among the pre-Columbian Indians"* (1938). Another work published at the time is the one by Maryon; *"Archaeology and Metallurgy: II. The Metallurgy of Gold and Platinum in Pre-Columbian Ecuador"* (1941). The collector Max Konanz published in 1944 his *"El arte entre los aborígenes de la provincia de Manabí"*.

It is, however, during the decades of 1950 and particularly 1960 when the archaeological explorations, and the metal finds, become frequent and noticeable. That is also when the collection of the *Banco Central*, that started in the 1940s gets bigger so that it can offer researchers a larger and more varied corpus of objects than had ever been available beforehand for research.

It is then that two works by Zevallos appeared: *"Tecnología metalúrgica arqueológica. La elaboración del alambre"* (1956) and *"Estudio regional de la orfebrería precolombina de Ecuador y su posible relación con las áreas vecinas"* (1965/66) both exploring technology and proposing a framework for native metallurgy in the South American environment. Bushnell: *"The archaeology of the Santa Elena peninsula in south-west Ecuador"* (1951) accounts some interesting finds. Also from this time is the book by Ross Christensen *"A Recent Excavation in Southern Coastal Ecuador"* (1954), a report of copper objects from the El Oro province.

However, the most influential work is that of Meggers, Evans and Estrada that defines several issues in the study of the archaeological context of metallurgy. Among their works we can cite, by Estrada, *"Ultimas civilizaciones prehistóricas de la cuenca del río Guayas"* (1957) and *"Prehistoria de Manabí."* (1957) a record of sites in the Manabi province and a partial typology of this industry in the central coast; by Evans and Meggers, *"Preliminary Report on Archaeological Investigations in the Guayas Basin, Ecuador"* (1954) in which the finds of the famous site of La Compañía are reported and the synthetic book by Meggers *"Ecuador"* (1966), part of the British series "Ancient Peoples and Places" that contains a chapter dedicated to the Milagro-Quevedo metallurgy.

Also important during these decades are the documents written by Olaf Holm, a researcher of the *Banco Central*, who unveils technological aspects of metallurgy in his article *"Copper needles from Manabí, Ecuador"* (1963), characterises problematic types in *"Money Axes from Ecuador"* (1966/67), interprets technological processes in *"Fuelles que son unos cañutos (Un comentario etno-arqueológico)"* (1968), proposes a global vision of the industry in *"Orfebrería Precolumbina del Ecuador"* (1970) and reveals the function of weapons in *"Lanzas Silbadoras"* (1977).

With the beginning of the decade of 1970 a new generation of scientists took over the thematic area, even though some pioneers kept on working in the field and with museum collections. Two of the new researchers were McCall and Bucheitt who in 1971 presented *"Metallographic Studies of Archaeological Artifacts from Ecuador"*, a metallographic study focused on the artefacts excavated at the site of La Compañía, Guayas. Warwick Bray approached the subject from an archaeological point of view relating Ecuador with the rest of South America in *"Los antiguos artífices americanos"* (1971).

In southern Colombia, by that time, there were both the scientific finds of the French team headed by Jean François Bouchard in the region of Tumaco and the wave of treasure hunting in the Ipiales plateau. Bouchard published *"Hilos de oro martillado hallados en la Costa Pacífica del Sur de Colombia"* (1979) and later *"Las más antiguas culturas precolombinas del Pacífico ecuatorial septentrional"* (1986). Some objects rescued from treasure hunting were classified and studied by Clemencia Plazas in *"Orfebrería prehistórica del altiplano nariñense, Colombia"* (1977/78). Later on Bouchard wrote again on the archaeology of the coastal region referring to the context of metallurgy in *"Arqueología de la costa del Pacífico nor-ecuatorial. Evaluación preliminar de los cambios ocurridos en los últimos decenios"* (1995) The results of these works are important for the knowledge of the Ecuadorian metallurgy of La Tolita and Carchi respectively.

Douglas Ubelaker excavated in Manabi at the beginning of the 1980s and in *"The Ayalan cemetery: A late intermediate period burial site on the south coast of Ecuador"* (1981) reports burials with metallic offerings associated to several C14 dates. The excavations of Marie Sutliff, in Guayas, recover copper artefacts in different stages of manufacture; their study allowed her to propose the existence of a domestic metallurgical production not oriented towards elite consumption. She also reconstructed the entire operating chain of manufacture of copper objects. Her thesis, *"El proceso productivo metalúrgico de la cultura Milagro: el caso de Peñón del Río"* (n.d.) has not been published yet but there are two related articles: *"Domestic Production of Small copper artefacts during the Milagro occupation al Peñón del Rió (Guayas Basin)"* y *"Contextual evidence for non – elite metallurgical production and use in Milagro Society"*, both from 1989.

The archaeo-metallurgist David Scott finished in 1982 his extensive doctoral thesis on metallurgy from northern Ecuador and southern Colombia; this work, as yet unpublished, which contains numerous and precise metallographic and compositional studies, is titled *"Pre-Hispanic Colombian Metallurgy: Studies of Some Gold and Platinum Alloys"*. Scott published the following articles related to his thesis: *"Depletion Gilding and Surface Treatment of Gold Alloys from the Nariño Area of Ancient Colombia"*; (1983) *"Dorado por Fusión y Dorado de Lámina en Colombia y Ecuador Prehispánicos"* (1985), as well as: *"Ancient Platinum Technology in South America"* (1980) and *"Pre-Hispanic Platinum Alloys: Their Composition and use in Ecuador and Colombia"* (1994) both wrote together with Warwick Bray. *"Orfebrería prehispánica de las llanuras del Pacífico de Ecuador y Colombia"* (1988) and *"Pre-Hispanic Platinum Alloys: their Composition and Utilisation in Ecuador and Colombia"* (n.d.) were written by Scott and Jean François Bouchard; finally *"La soldadura con aleaciones de oro en la América antigua: un análisis de dos pequeños adornos provenientes del Ecuador"* (1990) was written by Scott and E. Doehne. This series of articles by Scott and his collaborators explore in depth the theme of platinum technology and provide information about soldering, fusion gilding, gold, tombac and copper plating.

Heather Lechtman's work published in 1986 as *"Traditions and Styles in Central Andean Metalworking"* focused, as most of her publications on the Central Andes, has some important references to Ecuadorian metallurgy. Together with Dorothy Hosler and Olaf Holm, Lechtman carried out an extensive survey about money axes that helps them establish the relationship between the coastal Ecuadorian metallurgy and that of western Mexico (States of Colima, Jalisco and Nayarit). The result was published as *"Axe-Monies and their relatives"* in 1990; a synthetic version with some additional data was published by Hosler in 1998 as *"Los orígenes andinos de la metalurgia del occidente de México"*.

The results of the excavations of the Spanish Mission to the site of Ingapirca were the subject of a preliminary publication in 1975 by José Alcina in *"Excavaciones arqueológicas en Ingapirca (Ecuador)"*; Antonio Fresco, published the detailed report as *"La arqueología de Ingapirca (Ecuador)"* in 1984. In this site most of the pre-Inca graves had copper objects that were examined by Andrés Escalera and Maria Angeles Barruiso; the technical report *"Estudio científico de los objetos de metal de Ingapirca (Ecuador)"* came out in 1978. The same mission excavated in Esmeraldas and reported its finds in *"La arqueología de Esmeraldas, Ecuador"* by José Alcina (1979); in this region there were also metallic finds associated with late occupations.

Donald Collier and John Murra conducted a series of excavations in the province of Azuay, near Ingapirca, finding several copper objects described in *"Reconocimiento y excavaciones en el sur andino del Ecuador."* (1982). This decade closed with two articles by Francisco Valdez; the first one discusses the role of gold work in the La

Tolita community: *"Proyecto Arqueológico La Tolita"* (1987), which is also a catalogue for the exhibition of the results of archaeological work in the area. The second article titled *"Le Symbolisme du Naturel et du Social"* (1989) is part of another catalogue for an international exhibition of Ecuadorian archaeology.

If we are to judge on the basis of the number of bibliographical references found, the decade of 1990 is marked by a growing interest in the Pre-Columbian metallurgy of Ecuador. We can mention the short article by Robert Liu *"Precolumbian personal adornment. La Tolita / Tumaco and Jama Coaque"* published in 1992 that contains descriptions and aesthetic considerations about the mentioned regional groups. Jorge Marcos, one of the most prominent persons in the Latin American social archaeology, published *"El Mullo y el Pututo. La articulación de la ideología y el tráfico a larga distancia en la formación del Estado Huancavilca"* in 1995 and *"A Reassessment of the chronology of the Ecuadorian Formative"* in 1998; both contain statements as to the role of metallurgy in the regional exchange and in the formation of hierarchical social structures of the Ecuadorian coast.

The almost encyclopaedic doctoral thesis of Salvador Rovira approaches the technological aspects of regional groups of northern Ecuador and southern Colombia and characterises them on the basis of that condition, as well as their typology and chronology. *"La metalurgia americana: análisis tecnológico de materiales prehispánicos y coloniales"* has not yet been published. Eugen Mayer published in 1992 his book *"Armas y herramientas de metal prehispánicas en Ecuador"*, which is part of a series on weapons and tools in different regions of South America; this volume analyses in detail a large collection of these objects and offers a typology of bronzes and coppers in this zone.

In the southern coast of Colombia Diógenes Patiño conducted digs from 1993; in that year he reported finding metal objects associated to very early C14 dates in an excavation report titled *"Más evidencias sobre orfebrería temprana en Tumaco-Tolita, Costa Pacífica"*. Later on he supplemented the previous information in: *"Arqueología y metalurgia en la Costa Pacífica de Colombia y Ecuador"* (1997) and *"Orfebrería prehispánica en la Costa Pacífica de Colombia y Ecuador: "Tumaco-La Tolita"* (1998). An important conclusion of his work was that the occupation of the La Tolita-Tumaco culture extended towards the central Pacific coast of Colombia. This was confirmed by David Stemper and Hector Salgado in their article *"Metalurgia prehispánica y colonial-republicana en el Pacífico colombiano"* (1993), in which they reported evidences of the subsistence of the metal working tradition in the region of Buenaventura, Colombia, even up to modern times. In another article Salgado, Stemper and Florez discussed in depth the history of the site of La Bocana providing new C14 dates: *"Sociedades complejas en el litoral Pacífico: fragmentos de historia reconsiderados desde La Bocana."* (1995).

Another work by David Stemper analysed the role of metallic objects in the persistence of the chiefdoms of the River Daule. His book *"The persistence of Pre-Hispanic Chiefdoms on the Río Daule, Coastal Ecuador"* (1993) brings together the analysis of metal artefacts with a complex environmental study. In 1994 the article written by Thilo Rehren and Mathilde Themme *"Pre-Columbian Gold Processing at Putushio, South Ecuador: The Archaeometallurgical Evidence"* was published. This document exposes one of the most interesting gold working contexts in the province of Loja, southern Ecuador, associated to one of the earliest C14 dates for metallurgy in South America.

Another project that resulted in the recovery of metal objects was conducted by Markus Reindel and Nicolas Guillaume-Gentil; their article titled *"El proyecto arqueológico La Cadena. Estudios sobre la secuencia cultural de la cuenca del río Guayas"* (1995) sheds light about the use of metal objects in the Milagro society. In the north coast the joint work of French and Spanish archaeologists provides some more information on metallurgy; this is the case with the article of Mercedes Guinea *"La metalurgia del cobre en la costa norte del Ecuador durante el período de integración"* (1998) that refers to the manufacturing of metal objects in the Late Pre-Hispanic Period. Inca metallurgy, mostly ignored in Ecuador, was summarised by Albert Meyers: *"Los Incas en el Ecuador. Análisis de los restos materiales, I y II"* published in 1998, has a good compilation of finds and contexts.

Back in southern Colombia we can mention the finds of Pablo Casas in the island of Gorgona, compiled in *"La Gorgona en tiempos precolombinos"* (1991). The studies on the technology of platinum and fusion gilding carried out at the time are noteworthy; among them there is the report by Bustamante, Garzón, Bernal and Hernández: *"Tecnología del platino en la fabricación de piezas de orfebrería precolombina"* (2007) and the one by Emilia Cortes centred on a gilded crown from southern Colombia: *"Tecnología y conservación de un ornamento prehispánico para la cabeza procedente de Nariño, Colombia"* (1997).

In the *Banco Central del Ecuador*, Patricia Estévez advanced in the study of technology as became evident in her articles: *"Estudio de objetos metálicos del sitio arqueológico La Tolita"* (1994); *"Oro y platino en la orfebrería prehispánica del Ecuador"* (1995); *"Platino en el Ecuador precolombino"* (1998) and *"Museo de Esmeraldas. Informe de análisis de objetos de oro"* (1999), as well as: *"The Technology of Early Platinum Plating: a Gold Mask of the La Tolita Culture"* (2002) written with Meeks and La Niece. All together these articles constitute an important effort towards the understanding of native metallurgical technology backed up by analyses and experimentation. Alfredo García, Carolina Jervis and Pablo López proposed the typology and variations of gilding in *"El dorado de las narigueras en el Ecuador precolombino"* (2000).

Two aspects of metallurgy that are indispensable for its understanding were approached respectively by Karen Stothert and Sara Rodicio García with Angel Riesco. The first one is reported in *"Fundición tradicional campesina en la costa del Ecuador"* by Stothert (1997), an ethnographic study of lost wax casting among the peasants of the Santa Elena peninsula; probably a technology of Pre-Hispanic origin. In the second case Rodicio and Riesco explore the ethno historic chronicles about the god mines of the southern Sierra, emphasising on their probable Pre-Hispanic exploitation: *"Minas de oro "Santa Bárbara" en Los Cañaris"* (1998).

We cannot close the relation of the studies of this decade without mentioning the texts written for the catalogues of museums and temporary exhibitions of metallurgy that most times are valuable synthesis of information that would otherwise be left disperse. In this category we have the catalogues of the museums and exhibitions of the *Banco Central*: *"Sala del Oro. Museo Nacional del Banco Central del Ecuador"* (1995) by Juan Fernando Pérez et al; *"Museo del Banco Central del Ecuador, Ibarra"* (1998) by Fanny Cisneros et al; *"Museo Regional de Esmeraldas"* by Alexandra Yepez (2000); *"Museo del Banco Central del Ecuador, Riobamba"* (2002) by Santiago Ontaneda and Antonio Fresco, and *"Signos Amerindios. 5.000 años de arte precolombino en el Ecuador"* (1992) by Francisco Valdez and Diego Veintimilla.

The best known episode of a scientific debate in the history of the study of Pre-Columbian metallurgy of Ecuador occurred between the end of the 20th century and the beginning of the 21st century. The discussion about the "Golden Suns" of Ecuador had been fermenting in the academic circles as pure gossip, but it exploded with Karen Bruhns´s *"Huaquería, procedencia y fantasía: los soles de oro del Ecuador"* published in 1998 but widely known since the previous year. The reply by Constanza di Capua saw the light even before the publication of Bruhns. The article *"Una atribución cultural controvertida"* (1997) controverted the statements disseminated by Bruhns. This, in turn, gave rise to *"Estudio analítico de la mascara funeraria: Sol, cefalo, antropo, motivo zoo"* by Patricia Estévez (2002) that never got published and also to the project titled *"Proyecto de Metalurgia Prehispánica"* (2002) by Estévez, Fresco, Valdez and Yépez that was not made public. The discussion was finally closed at the 51st International Congress of Americanists (2003) with a paper by Valdez, Estévez and Barrandon: *"Mucho ruido y pocas nueces. El epílogo de la controversia del origen de los soles de oro del Ecuador"*. (2007)

With the advent of the 21st century, and as a consequence of the research and debates that took place in the past decade, new projects and interpretations arose. The discussion of the "Golden Suns" gave way to a very well structured report: *"Identificación mineralogica de las fuentes del oro precolombino utilizado en la metalurgia prehispánica del Ecuador"* by Barrandon, Valdez and Estévez (2002). The digs of Leon Doyon in the site of Florida, north of Quito, recorded several graves with metallic offerings associated to C14 dates; the report was titled *"Apuntes hacia un nuevo entendimiento de la historia cultural del area Carchi-Nariño"* (2002).

In Colombia Roberto Lleras and Luz Alba Gómez re-examined the metallurgy of the southern Andes together with that of the northern Andes of Ecuador in succesive articles: *"The Influence of Central Andean Metallurgy in the Highlands of Southern Colombia"* (1999); *"La Problemática de la Arqueología Nariñense Vista desde la Metalurgia"* (1999); *"El Tiempo en los andes del norte de Ecuador y sur de Colombia: un Análisis de la Cronología a la Luz de Nuevos Datos"* (together with Javier Gutiérrez, in 2007); *"Desarrollo y simbolismo dual de la metalurgia de Nariño y Carchi"* (only Gómez, 2003) and *"La Problemática del Capuli, Piartal, Tuza: Una nueva clasificación orfebre"* (2006). Garzón, Bernal and Hernández contributed to the revision of the metallurgical classification of the southern Andes with *"Nariño, algunos desarrollos tecnológicos en su orfebrería"* (2007).

We should also mention the most recent summary of South American metallurgy written by Warwick Bray: *"Metal Artefacts in the American World: Archaeological Evidence"* (2000), the unpublished synthesis by Sandra Mendoza *"Las Gentes y el Oro en la Costa Pacifica Sur"* (2000) written for the new layout of the *Museo del Oro de Colombia* and the interpretative article by Paulina Ledergerber *"Ecuador: Uno con el sol y la luna"* (2004) which is an interesting essay on the symbology of metallurgy. There are recent documents dealing with technology; one by Patricia Estévez *"Legado tecnológico en orfebrería"* (2003), another by Salvador Rovira *"Un fragmento de placa dorada precolombina de Ecuador: estudio analítico"* (2004) and the article by Noguez et al *"About the Pre-Hispanic Au-Pt "sintering" technique for making alloys"* (2007). Francisco Valdez and his co-workers found in Esmeraldas a context with metal objects dated between 780 and 918 B.C. that are interpreted as an initial phase of the La Tolita-Tumaco metallurgy: *"Evidencia temprana de metalurgia en la costa Pacifica ecuatorial"* (2007).

A provisional balance of the studies of the Pre-Columbian metallurgy of Ecuador suggests that:

a) The academic interest in metals has varied along the last century. After Rivet, Verneau, Arsandaux, Jijon y Caamaño and other pioneers called attention upon metallurgy several decades passed during which the archaeologists ignored this vital aspect of Pre-Hispanic cultures. Then the interest surged again and some researchers, equipped with new tools for instrumental analyses and renewed theoretical approaches undertook the study producing very good results.

b) In spite of this there have been frequent finds of metallic objects in archaeological digs and, in these cases, the artefacts have been analysed from the typological point of view and also using metallurgic and metallographic techniques.

c) Collection studies have not been too frequent and, when they occur they do not tend to involve large numbers of objects. Nevertheless, the importance of de-contextualised artefacts has been proven, especially for the study of technology.

d) In most cases, both in the past and presently, there is a strong tendency to emphasise that there have been external, especially Central Andean, influences determining the Ecuadorian metallurgy. This has led to a diminished interest in the study of regional developments and the role that the territory of Ecuador played in the origin and development of metallurgy in America.

e) It is evident that among the archaeologists that have dealt with the theme from the 1970s onwards there is a general consensus about the general framework of the Pre-Hispanic metallurgy. Nevertheless, aside from the exhibitions of the *Banco Central* and their catalogues, the consensus has not been explicitly expressed in a general and detailed text.

This is, in summary, the history and results of the study of archaeometallurgy in Ecuador. Metallic objects have been known for almost 500 years and studied along the last hundred years. In this lapse our perception of metallurgy has changed markedly to the point that we now regard it as part of our national heritage. Since it was discovered and plundered by the European conquerors in the 16th century, until the time came when its scientific study, conservation and museographic exhibition became the rule, metallurgy has continued marvelling us; we are sure, though, that even if we explore it thoroughly there will always be more to learn.

FIGURE 3 OLAF HOLM, ONE OF THE PIONEERS OF THE STUDY OF METAL ARTEFACTS IN ECUADOR

Chapter 3
Metallogenesis and metal resources in Ecuador

The Andean Cordillera, that crosses the territory of Ecuador from north to south, is recognised from the metallogenic point of view as a geological formation rich in metallic minerals and native metals, especially copper, gold, silver and arsenic, among those of interest for the Pre-Hispanic epoch. In general terms metal deposits are associated to magmatic activity areas, even though plaque tectonics also determines the production and placing of magmas and the existence of flux channels (Oyarzún 2000).

It follows that the origin of metal deposits is associated to the active segments of strato-volcanoes, as the one located in Ecuadorian territory between 1° north and 5° south, and with traverse faults as the one that gives rise to the deposit of porfidic copper in Chaucha, Ecuador (Oyarzún 2000).

Specialists recognise the existence of metallic provinces; among them the provinces of copper, gold and silver and the poli-metallic are of special interest with regard to Pre-Columbian metallurgy. The first one covers almost all of the Andean Cordillera, even though the characteristics and sizes of the deposits are dissimilar (Oyarzún, 2000). In Ecuador there are deposits of porfidic copper and vein deposits disseminate throughout the Sierra (i.e. Chaucha); some were probably exploited since Pre-Hispanic times, even though there are no archaeological or documentary evidences of such activity (Hosler 1990, Guinea 1998, Holm 1970).

Even though most of these deposits are not commercially viable at present, due to their low productivity, it is possible that they were an adequate source of raw material for the moderate levels of consumption of Pre-Columbian times. The metal smiths of the Sierra (Carchi-Nariño, Puruha, Cañari) must have had at their disposal local deposits of copper. In that sense, and in spite of the lack of information, the supply of copper in this region is not a difficult problem.

On the contrary, in the coast, a region where there flourished an important copper metallurgy, there are no deposits of this metal, which must have been imported. The only references we have to contradict what we have just stated come from the Jaboncillo Hill in Manabí, where presumably copper was extracted by means of deep underground galleries (Jijon y Caamaño in Rivet et Arsandaux 1946) and the possible existence of melting furnaces in Manabí cited as *"...the only vestige we have about real Pre-Columbian mining in Ecuador"* (Holm 1970). Taking into account that these two brief references have not been confirmed and are not consistent with what we know about metallogenesis we must consider that metal smiths from Manteño-Huancavilca,

Milagro-Quevedo and La Tolita-Tumaco must have obtained copper from not too distant mines of the Sierra, probably located in Bolívar, Azuay, Cañar, Imbabura or Carchi (Hosler 1990). Other authors (Guinea 1998) maintain that copper might have been carried from much more distant deposits in central and northern Peru.

We do not know either if the copper that arrived in the coast, from the Sierra or Peru, had already been subject to some degree of processing. In a coastal site (Peñón del Río) where evidences of manufacturing of copper objects have been found there are no traces of the process of refining, thus indicating that manufacturing in there started with metallic copper (Sutliff 1989, 1990). In this case the processes of grinding, refining and melting of the minerals must have occurred near the mines and the copper should have been taken to the coast in the form of ingots. We are not sure, though, that this was the general case.

Guinea (1998) believes that the metal must have been transported in the shape of spherical nodules (*prills*) that come out of the initial smelting process or in the shape of ingots (cards, axes) already refined. The metal smiths in Atacames would have obtained the material through the Huancavilca trading net and would have used it to produce finished objects. To support this idea she cites *"...the strong technological and instrumental similarity with the Peruvian materials from Middle Sican..."* and the contemporaneity of both industries (Guinea 1998)

There are many compositional analyses of copper and arsenical coppers found in archaeological digs and in some cases it has been possible to determine the unit of provenance of raw materials for assemblages found in different graves (Escalera y Barruiso 1978). The same authors, as well as Jijon y Caamaño (1929), Bergsoe (1938), Reichlen (1942), Rivet y Arsandaux (1946), McCall y Buchheit (1971), Scott (1980), Rovira (1992) and Stemper (1993) cite the results of compositional analyses that reveal the fact that Ecuadorian copper contains as impurities small and variable quantities of silver, gold, nickel, antimony, cobalt, zinc, lead and iron.

Arsenic appears also as an impurity in Ecuadorian copper but, in certain objects, it is present in larger quantities. This seems to indicate that it was intentionally added, especially because when arsenic is present as an impurity in small quantities it tends to get lost in the processes of copper refinement. These data suggests the presence of an interesting pattern of natural alloying characteristic of Ecuadorian coppers that would differentiate them from those coming from elsewhere. We still need a particular study oriented towards the identification of the sources of raw material, a work that would require carrying out trace element or isotopic analyses of copper minerals from the Sierra deposits and its comparison with the same indexes of the archaeological objects.

The metallic province of gold and silver covers the whole mountainous zone of Ecuador. In it gold appears within different geological formations and alloyed with several metals. From the metallogenic point of view these deposits belong to the type of epithermal veins (Portovelo, Pilzhum and Molleturo), *skarn* (Nambija y Pachicutza), *stockwork vein* (Chinapitza), intrusive *breccia* (Gaby) as well as those associated with porfidic copper (Fierro Urco) (Gemutz et al 1992 in Oyarzún 2000). Paladines y Díaz (in Estévez 2003) recognise seven mining districts: Portovelo-Zaruma; Ponce-Enríquez; Nambija; Chinapintza; Macuchi-La Plata; Molleturo and Azuay. Each district has different characteristics with respect to its productivity of gold and silver that fluctuate between 1 and 80 gs of gold and 24 to 392 gs of silver per ton of parental material. In addition, some of them contain up to 5% of copper.

Many of the rivers that run down from the Sierra wash the gold veins and carry weathered gold that is deposited in the sand beaches of those rivers in the coastal plains, where the speed of the water flow decreases. The Cordillera Real and the rivers that flow from it are recognised as an important source of alluvial gold (Estévez 2003). This meant that gold and its sub-product, silver were available locally in most of the Sierra and Costa. However, the quantity, accessibility and quality of the deposits varies greatly from region to region.

It is possible that the vast majority of the mining activity in Pre-Hispanic Ecuador was limited to washing alluvial sands in the coastal area; this mode of mining is technically simpler and has the advantage of providing gold that has been naturally refined, because most natural impurities are lost during the washing and weathering. In the Coast there are remarkable differences between the deposits. In the north, province of Esmeraldas, most deposits correspond to platinum gold (Scott y Bray 1980); those deposits are an extension of the deposits of the Colombian Pacific coast that cover the Ecuadorian Western Cordillera as far south as the valley of the Guayllabamba. In the central and southern coast the placers are mainly gold and silver gold (Oyarzún 2000).

Up to date no documents have been found with concrete information about placer mining in the Ecuadorian coast or in southern Colombia. Oyarzún (2000) mentions the early colonial exploitation of the Portovelo deposits which might have started in Pre-Columbian times. Estévez (2003) reports several sites (without photographic records) in which there are vestiges of channels, possibly made by the Indians, used to conduct river water to gold washing facilities. The same author affirms that in the same sites there are structures for grinding and carrying ground material.

In the Sierra there are deposits known since very early colonial times, such as the famous mines of Santa Bárbara in Cañar (Rodicio y Riesco 1998). It is evident that there was a Pre-Columbian mining tradition in this area because as early as 1539, even before the colonial authorities were established, the White neighbours were already exploiting them. It is likely that the large quantity of gold that was found in the graves of Cañar was extracted from the lagoons of Santa Bárbara and other neighbouring mines (Rodicio y

Riesco 1998; Uhle 1922). Estévez (2003) believes that the terraces of the Shincata and Betas rivers were exploited before the Incas. Such rich mines are not to be found all along the Sierra, thus leaving us the question of the provenance of gold used in Carchi-Nariño and Chimborazo, among other areas.

The controversy of the Golden Suns of Ecuador helped us understand that it was possible to distinguish clearly between the placer gold of the north coast and the mine gold from the Sierra (Valdez, Estévez y Barrandon 2003). The platinum gold of Esmeraldas and Nariño (southern Colombia) appears in association with pre-Tertiary intrusions of maphic and ultra-maphic rocks of the Cordillera that underwent severe weathering processes (Scott y Bray 1994). Even though it is feasible to extract platinum in those deposits, the analyses carried out on over 150 objects led the authors to conclude that the platinum used in Pre-Hispanic times came from placers where it appears mixed with gold (Scott y Bray 1994).

Platinum gold is a material that entails difficulties for the metal smith. Besides platinum this type of gold usually has palladium, rhodium, iridium, osmium, ruthenium, iron and small quantities of copper. The metals of the platinum family and iron render special hardness to this gold, makes alloying with copper to make tombacs problematic, makes it difficult to work and brittle when hammered (Scott y Bray 1994). The available methods in Pre-Columbian times were unable to ensure an effective separation of gold from the other metals; thus forcing the development of a technology fit to deal with this kind of mixtures of metals, as it effectively occurred in La Tolita-Tumaco. Nevertheless, whenever the desired alloy was a tombac they had to find gold from other sources without platinoids.

The silver used in Pre-Columbian metallurgy, both in the Sierra and Costa, is from the metallogenic point of view, a sub-product of gold because in Ecuador there are no deposits containing this metal (silver) by itself (Oyarzún, 2000). Up to date there are no studies explaining how the natives separated silver from gold. The analyses made (Rivet y Arsandaux 1946, Stemper 1993, DTI 2006) indicate that there are few objects of pure silver and that most of them are silver-copper alloys. It is not clear, though, if in every case alloying was intentional because it is also possible that metal smiths used silver minerals that contained copper as a natural impurity.

The small quantities of lead used in Pre-Columbian metallurgy (Rivet y Arsandaux 1946, Ubelaker 1981) might correspond to sub-products of copper or silver refining from minerals that contain it in small quantities. Rivet and Arsandaux (1946) propose that refining galena might render considerable quantities of lead that was used in Pre-Columbian times much more frequently than usually thought. From the metallogenic point of view lead is to be found in the deposits of the poli-metallic province (Oyarzún 2000).

When analysing the metallogenesis of Ecuador and the distribution of the deposits and prospects presently located, as indicative of the availability of metals for the Pre-Columbian metal smiths, it is important to take into account the huge difference between modern mining technology and that of the past. Most metal deposits recently discovered can only be exploited by removing and processing thousands of tons of parental material, others are at depths that can only be reached by means of powerful elevating machines and ventilation. Finally, some minerals require mechanical or chemical treatments that have been developed only in modern times. The range of deposits of copper, gold, silver, platinum, arsenic and lead available for Pre-Hispanic Indians was much more limited. In order to have a precise idea of what was the range of available deposits where Pre-Columbian mining could have been practised it is necessary to classify them according to criteria of accessibility and workability.

Even though the authors cited (Holm 1970, Hosler et al 1990) insist in that there are no references in Spanish chronicles about Indian mining, it is evident that ethnohistoric documents cannot be discarded altogether as sources of information. Frequently colonial documents other than chronicles such as *Visitas* (inspections), *tasaciones* (census), lawsuits and testaments contain useful information about economic activities in the first decades of the colonial era (Rodicio y Riesco 1998). Colonial mining was, as a general rule, a continuation of the Pre-Hispanic practices, albeit the use of new techniques and organisational models. The panorama of colonial mining in the 16th century and the first half of the 17th century must be very close to that of the Pre-Hispanic period.

Crossing the ethnohistoric information of the early colonial period with the data of metallogensis and filtering the results with respect to the accessibility of the deposits would play a double role. First it would allow us to establish a comparison with the provenances of objects from museum collections; we would then have some sort of correlation between mining sites and places where objects were used and deposited. But, this would also serve as the starting point for a field archaeology project aimed at finding and excavating ancient mining sites; i the course of such project probably many questions related to technology would be answered and we would have the basis to understand sources and exchange networks of raw material.

Figure 4 Chimborazo the highest strato-volcano in Ecuador; metal deposits are associated to volcanic activity.

FIGURE 5 ALLUVIAL RIVER PLACERS LIKE THIS ONE IN THE LOWLANDS OF THE PACIFIC COAST ABOUND IN GOLD AND PLATINUM.

Chapter 4

Early finds and the Initial Period

Even though, as previously stated, many researchers tend to consider that Ecuadorian metallurgy resulted from diffusion or cultural influences from the Peruvian Central Andes, the data gathered seems to point in another direction. What we are actually finding is that the territory of Ecuador might have been one of the regions where early metallurgical experimentation occurred in South America.

The first find indicating that metallurgy in Ecuador is very old was made by Mathilde Themme (Rehren y Themme 1994) in the site of Putushio, Loja province, south Sierra. There, in stratigraphic levels described as corresponding to Early Formative, Themme found several gold objects, among them tiny gold spheres adhered to a clay mould made for their manufacture. Themme's finds that belong to a metal workshop are associated to an initial C14 date of 1470 B.C. According to the author the metallurgical activity continues uninterruptedly thorough the 13th and 9th centuries (C14 dates, 1260 and 865 B.C.) even until the latest periods of the chrono-stratigraphic sequence (A.D. 1305). In this site Themme excavated a central circular depression lined with clay that must have served as a furnace. There are also evidences of cold working (Rehren y Themme, 1994).

Hosler (1997) briefly described the results of her excavations in Salango, province of Manabí, Central coast, where she said to have found objects *of "...gold, silver and copper, and hammered."* (Hosler, 1997). In the published texts there are no details about contexts, dates or associations, even though according to Hosler the objects belong to the Chorrera Phase and can be dated to around 1500 N.C., a date that in this region would correspond rather to the Machalilla phase. The same author claims that there are other later finds from Salango and the sites of El Azúcar and Cerro Alto in the Santa Elena peninsula, province of Guayas, namely copper needles manufactured prior to A.D. 500 (Hosler 1997). In the published text there are no exact data or laboratory numbers thus making it impossible to include them in a C14 dates database or correlate them to other comparable finds.

Zevallos (1965) excavated a grave in the site of Los Cerritos, in the seashore of the bay of Santa Elena, where he found a fragmented copper sheet and two tools for metalworking, part of a stone anvil and a hammer. The finds are associated to pottery of the Chorrera phase and a C14 date of 890 B.C. (Radiocarbon Database, 2006).

The most recent find related to this topic was made by the French-Ecuadorian team led by Francisco Valdez (Valdez et al 2007). In the site of Las Balsas, to the north of the province of Esmeraldas, north coast, in strata corresponding to the Tachina phase, Late Formative, a small sheet of a ternary alloy of gold-silver-copper made by hammering was found associated to two coincident calibrated C14 dates that place its deposition between 918 y 780 B.C. Even though it is just an isolated object, the context and association are precisely defined and unequivocally suggest the presence of relatively complex metallurgy in an area that would later become an important metal working focus. (Valdez et al 2007).

As yet there is no substantial corpus of C14 dates associated to metallic objects for Ecuadorian Pre-Columbian metallurgy. Such situation makes it difficult to establish relations or patterns of continuity between the Early Period and the developments that followed in three areas of the region of particular interest.

The first area of interest is the province of Loja, in the south Sierra, where the internal sequence of the site of Putushio by itself attests the permanence of metallurgical work in the area for around three thousand years (Rehren y Themme 1994). However the metallurgy of Putushio, even in its latest phases does not define a style tan can be linked to those that arose afterwards in the south Sierra Sur, such as Cañari and Puruha. What is evident is that the variety of techniques used at Putushio (Rehren y Themme 1994) is consistent with a technological nucleus that was important beyond the local scene; most probably the metallurgy of the south Sierra developed at Putushio and other similar sites that existed in the region. It is also probable that the Narrio phase might have included a metallurgical component. As in the case of the Chorrera phase, the metallurgy of Narrio must have been quite simple, both in formal and technological terms.

The second region of interest where some sort of continuity can be established is the north coast, closely related to the southern Colombian coast. The set of C14 dates for the region of distribution of the La Tolita-Tumaco Group (Initial Period and Early La Tolita-Tumaco) covers a span from the 8th century before our Era to the 2nd century A.D.(C14 dates: 720 B.C; 710 B.C; 470 B.C; 370 B.C; 325 B.C; A.D. 90 and A.D. 110). Even though the corpus of data is largely incomplete, in this area it is possible to track and locate in the time scale the initial phases of metallurgical experimentation, the consolidation of a regional style and the changes that followed along over eight hundred years.

The evidences provided by Hosler (1997) are much too succinct so as to evaluate the meaning of her finds and their relation with the regional metallurgy of the central coast. Nothing is said with relation to the persistence of metallurgical activity in the site of Salango after 1500 B.C. Nothing specific is stated about the finds of Santa Elena dated before A.D. 500 (Hosler, 1997). On the contrary the date supplied by Zevallos (1965) is extremely interesting, in spite of the scarce metallurgical evidences. There is, however, a wide chronological hiatus between this date (9th century B.C.) and the earliest available dates for the Regional Groups; a 2nd century date (A.D. 120) for Milagro-Quevedo (Stemper 1993) and a 7th century date (A.D. 730) for Manteño-Huancavilca (Ubelaker 1981). There is a gap of over a thousand years that obscures our understanding of the process of consolidation of metallurgy in the central coast.

Nevertheless, in spite of the deficiency of the data provided by Hosler (1997) it seems to be clear that there is some sort of association between vestiges of metallurgical work and pottery of the Middle Formative; the find of Zevallos (1965) confirms this idea. Even though it has been traditionally accepted that metallurgy appeared in Ecuador in the Late Formative, after Chorrera and Narrio, there are multiple evidences of metal production in the Middle Formative. The obscure period of over 1000 years might correspond to the development of Chorrera and Narrio metallurgies.

We would be compelled to ask then: Where are the Chorrera and Narrio objects? Even though in the museum collections there are no objects classified as Chorrera or Narrio it is quite feasible that they might be there classified under wrong headings or as undefined cultural attribution (Estévez, pers. com.). Most objects belonging to these metallurgical groups can be relatively simple and might, therefore, pass unnoticed or be allocated to other categories. Without composition analyses and microscopic examination it is very difficult to establish, for example, if any given simple hammered wire ring belongs to one or other regional group, much less if there is no information on the find context. This topic is far from being solved, but it is a key aspect in the history of Pre-Columbian metallurgy in Ecuador and it requires further research.

In the other regions of Ecuador there are no early evidences related to metal working. The sequence of Ingapirca (for Cañari) starts at A.D. 990 (Fresco, 1984); for Puruha there is a date from a museum object for A.D. 210 (Museo del Oro, 2005) and for Carchi-Nariño there is a solid corpus of dates starting at the site of La Florida in A.D. 130 (Doyon 1995). In general, and with the remarkable exceptions of La Tolita – Tumaco and Putushio, there is a dark period of over a thousand years during which the basic metallurgical skills must have extended across the Costa and Sierra and the regional styles gradually took shape. This period, characterised by the rise of powerful chiefdoms, regional exchange and the development of complex iconographic and symbolic systems, probably saw the emergence of true metal smiths and the proliferation of technological developments.

The following table summarises the information about C14 dates associated to metallic objects of the Initial Period:

Date	Canton	Laboratory number	Bibliographical reference
1470 ± 255 B.C.	L-Loja	Hv-16798	Rehren y Themme 1994

Date	Canton	Laboratory number	Bibliographical reference
1260 ± 180 B.C.	L-Loja	Hv-16799	Rehren y Themme 1994
890 ± 90 B.C.	G-S. Elena	Wiss - 115	Zevallos 1965
865 ± 95 B.C.	L-Loja	Hv-16797	Rehren y Themme 1994
720 ± 35 B.C.	E-La Tolita	Gif-11900	Valdez et al. 2006
710 ± 60 B.C.	E-La Tolita	Beta 181458	Valdez et al. 2006

Chapter 5

Great Regional Groups: La Tolita –Tumaco

The La Tolita – Tumaco regional group is one of the most abundant in the collection of the Ministry of Culture. As stated in a previous chapter this area has also been the subject of a particularly intensive attention on the part of researchers; these two factors have yielded a good deal of information thus allowing us to propose several hypothesis regarding technological and chronological aspects. There are also in the collection of the *Museo del Oro de Colombia* a group of objects from the same area thus adding to the available information.

Geographic Distribution

The following table shows the quantitative and percentage distribution of the objects classified as part of the Regional Group La Tolita – Tumaco in the collection of the Ministry of Culture (Quito):

Province	Canton	Number of objects	Percentage of total
Unknown	Unknown	9	0.5
Esmeraldas	Atacames	5	0.3
	Esmeraldas	216	12.4
	La Tolita	1,384	79.5
	Montalvo	1	0.05
	Quininde	2	0.1
	Rió Verde	12	0.7
Manabi	Convento	1	0.05
	Limones	1	0.05
	Mongoya	1	0.05
	Rocafuerte	17	1.0
	San Isidro	92	5.3
Total		1,741	100

The map displays the provenance sites of objects belonging to La Tolita – Tumaco (green circles) with relation to the general distribution of all objects of the collection of the Ministry of Culture (red circles):

The following supplementary table displays the distribution of metal objects of the La Tolita – Tumaco group in southern Colombia in the collection of the Museo del Oro, in quantitative and percentage terms:

Great Regional Groups: La Tolita – Tumaco

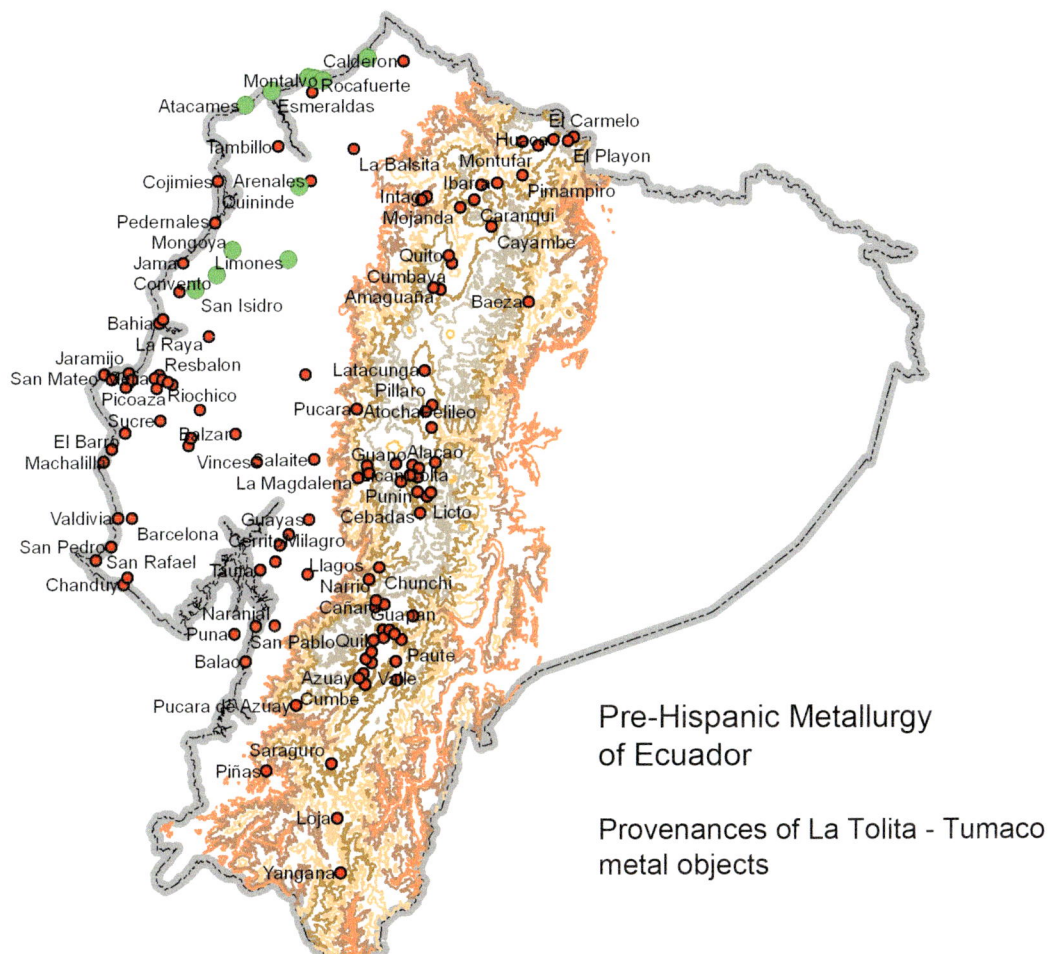

Pre-Hispanic Metallurgy of Ecuador

Provenances of La Tolita - Tumaco metal objects

Departmenát	Municipality	Number of objects	Percentage of total
Unknown	Unknown	46	21.9
Cauca	El Bordo	8	3.8
	Timbiqui	18	8.6
Nariño	Barbacoas	3	1.4
	Nariño undefined	9	4.3
	Tumaco	116	55.2
	Other municipalities	1	0.5
Valle	Buenaventura	1	0.5
	Restrepo	7	3.3
	Yotóco	1	0.5
	Total	210	100

The following map illustrates the provenances of La Tolita – Tumaco objects in southern Colombia:

Pre-Hispanic Metallurgy of Ecuador

Provenance of La Tolita - Tumaco metal objects in southern Colombia

What is immediately evident is that there is a great deal of concentration of finds, both in Ecuador and Colombia. In the case of Ecuador almost 80% of the objects come from the site of La Tolita, while in the case of Colombia over 55% come from the municipality of Tumaco. Even though La Tolita was for sure an exceptionally important ceremonial and burial site (Alcina 1979, Valdez 1987) it is unlikely that so many metal objects would be buried there because there were other ceremonial regional centres of some importance. It seems more likely that most finds made in many different sites of the region were grouped together and when sold were declared as found in La Tolita regardless of their real provenance.

Something similar might have happened in the Colombian side where finds made at the vast coastal plains or in the foothills were grouped and sold to the *Museo del Oro* as coming from Tumaco. Notwithstanding that Tumaco bay was also the site of an important ceremonial and burial place.

Apart from the two eponym sites (La Tolita and Tumaco) some objects belonging to this group were found at the seashore of Esmeraldas (Rió Verde, Montalvo, Esmeraldas and Atacames) and on the north coast of the province of Manabí (Convento, Mongoya and Rocafuerte). Going inland finds become increasingly scarce and are recorded only at Quininde and Limones. In Colombia finds are scarce in the plains north of Tumaco (Barbacoas, Timbiqui and El Bordo) but they are present from there up to the bay of Buenaventura. There is also a small group of objects that was carried across the Western Cordillera and was found at municipalities of the Cauca Valley (Restrepo and Yotóco).

The geographic distribution, in summary, seems to obey a pattern determined by a concentration in important ceremonial centres as La Tolita and Tumaco, combined with much scarcer finds in less important sites located near the seashore and finally a few objects buried inland.

Chronology

As explained earlier, the region has seen several important archaeological research projects on both sides of the border; this has yielded a series of 15 C14 dates summarised in the following table:

Date	Municipality	Laboratory Number	Period	Reference
470 ± 90 B.C.	N-Tumaco	Beta 82931	Early	Patiño 1993
370 ± 60 B.C.	N-Tumaco	Beta 82930	Early	Patiño 1993
325 ± 85 B.C.	N-Tumaco	Ny - 642	Early	Bouchard 1982
55 ± 35 B.C.	V-Buenaventura	Pitts 1145	Early	Salgado et al 1995
A.D. 90 ± 60	E-La Tolita	GIF-6815	Early	Scott y Bouchard 1988
A.D. 110 ± 60	C-El Bordo	Beta 20603	Early	Patiño 1988

Date	Municipality	Laboratory Number	Period	Reference
A.D. 770 ± 50	E-Atacames	CSIC-290	Late	Alcina 1979
A.D. 800 ± 60	E-Atacames	CSIC-289	Late	Alcina 1979
A.D. 850 ± 50	E-Atacames	CSIC-287	Late	Alcina 1979
A.D. 870 ± 50	E-Atacames	CSIC-286	Late	Alcina 1979
A.D. 875 ± 80	N-Tumaco	IAN-112	Late	Scott y Bouchard 1988
A.D. 880 ± 60	E-Atacames	CSIC-285	Late	Alcina 1979
A.D. 930 ± 50	E-Atacames	CSIC-280	Late	Alcina 1979
A.D. 1135 ± 35	V-Buenaventura	Pitts 1149	Late	Salgado y Stemper 1993
A.D. 1185 ± 50	V-Buenaventura	Pitts 1153	Late	Salgado y Stemper 1993

Most authors (Bouchard 1986, 1995; Valdez 1987) affirm that the La Tolita – Tumaco metallurgical tradition seems to have been well established by the 5th century before our Era. This estimate is consistent with the age generally accepted for the culture itself (ca. 500 B.C.). However, it is important to remember that the formative phase of this culture is present in the region from 900 B.C. (Valdez et al 2007).

The second important point to consider is that with respect to metallurgy, as is the case for the pottery and stratigraphic sequences (Bouchard 1995), there are two very well defined periods; the available data indicate that the first one (Early) goes from 500 B.C. to A.D. 200 approximately, while the second one (Late) is safely recorded from A.D. 700 and is still present at the time of Spanish conquest in ca. A.D. 1500 (Alcina 1979, Salgado y Stemper 1993). The time lapse between A.D. 200 and A.D. 700 has not been documented yet with respect to metallurgy; a fact that prevents us from precisely defining the transition between the Early and Late metallurgical phases. The evidence seems to point in the direction of an uninterrupted metallurgical activity in the region from 900 B.C. until A.D. 1500, but this is still to be confirmed.

The nature of the differences between the Early and Late periods of La Tolita – Tumaco metallurgy is addressed in the chapters dedicated to technology and typology. It is important to note that after the European occupation the native population continued making metal objects and that the African groups that arrived in the area brought with them metallurgical skills. Thus, in the colonial period there emerged a mestizo jewellery tradition that became particularly important in the Chota-Mira valley and in Barbacoas, northwest Nariño. In some archaeological sites there have been finds of metal objects that in view of their constituent material and their manufacturing techniques can be surely ascribed to the colonial or republican periods (Guinea 1998, Salgado y Stemper 1993).

Technology

In this section, as well as in all the other sections where technology is discussed, we will avoid describing the most common and well known metallurgical techniques.

Processes such as lost wax casting, hammering with alternate annealing, repousse and depletion gilding have been described many times and there are hundreds of references on them. We will, therefore, describe only complex techniques or those that, not being complex, are singular and particular to one or more of the Groups studied and are not used elsewhere. We will not describe in detail either the complete manufacturing processes of certain special objects because that would add many pages to the book; if needed those descriptions are available in the references cited.

The technology of the La Tolita – Tumaco Group has been the centre of attention of archaeologists and archaeo-metallurgists since the beginning of the 20th century; this is due to the presence of platinum, an attractive metal difficult to handle even nowadays. Nevertheless, the use of platinum is just one of the technological achievements of the La Tolita – Tumaco metallurgy that is characterised by the use of a large range of metals and alloys and a wide variety of techniques.

It would be nice to have a clear division between the objects belonging to the Early and Late periods of the region, so that the discussion on technology could explain the changes that undoubtedly occurred in the course of that transition. However the objects of the collection have not been classified in these two categories and that task was not feasible within this project. This means that there is a restriction that we need to acknowledge and that forces us to treat together objects that belong to different periods.

There are over 100 composition analyses for the La Tolita – Tumaco metallurgy based on XRF, Microprobe, ICP-MS, NAA and AAS and over 30 metallographic analyses (Barrandon, Valdez y Estévez 2002, Estévez 1999, Rovira 1990 and Arqueometalurgia 2005). This corpus of information is taken into account in this section to point out the general trends shown by the results obtained.

Matrices display a typical composition divided into two distinct groups: the first one is composed of sinters of silver gold with platinum and silver gold without platinum. Within this group copper appears as an impurity in low proportions (>6%). The other group is constituted by gold rich tombac, copper rich tombac and copper. Among the minor elements and trace elements there is palladium, osmium, ruthenium, nickel and zinc.

The visible microstructures of the first group of objects (gold with platinum) correspond to sinters with platinum rich phases (globules) surrounded by gold phases. In some cases there is a partial dilution of the platinum globules I the god phase due to hammering and annealing. In the metallographies of the objects of the second group (tombac and copper) there can be seen cold-working structures, melting with subsequent hammering and annealing, melting without further treatment, gilding and soldering by exudation.

In La Tolita – Tumaco alluvial gold was frequently used; in that region it is naturally mixed with variable proportions of silver, platinum and other metals of the platinum family, such as rhodium, palladium, iridium, osmium and ruthenium, besides iron. The composition analyses suggest that in some cases manual separation was applied in order to reduce the quantity of platinum and platinoids present because they render the sinter special characteristics that make it difficult to work (Scott y Bray 1994). These processes of separation are not completely efficient, not even those applied under modern laboratory conditions (Rivet 1946); in certain cases it would have been easier to preserve or even to increase the quantity of platinum naturally present, separating gold to obtain a powder rich in platinum and platinoid particles.

The melting techniques available at Pre-Hispanic times would not allow reaching temperatures beyond 1200 to 1400 °C (Scott y Bray, 1994) so that platinum having a melting point of 1770 °C, whenever present in substantial quantities, would remain in its solid phase in the melting processes and would appear as grains or inclusions. If, on the other hand, the mechanical separation had left only traces of platinum, it could dissolve in gold in the melting process resulting in a true alloy of yellowish white colour, harder than gold and liable to cold working (Bergsoe 1937-8).

Whether it was because it was not always feasible to separate mechanically gold from platinum or because it was considered desirable to take advantage of the colour properties of platinum, a set of technologies was developed that became known as sintering or compenetration. For this particular region it was initially proposed by Wolf (1879), then described in detail by Bergsoe (1937-38) and afterwards fully developed by Rivet (1946), Scott (1982), Scott y Bray (1980, 1994), Scott y Bouchard (1988, n.d.), Estévez (1994, 1995, 1998, 2003), Meeks et al (2002) Bustamante et al (2007) and Noguez et al (2007).

In summary the process consists of heating the mixture of gold and platinum until the molten gold forms a matrix that surrounds the platinum grains rendering strength to the material (Bergsoe 1937-38). The repeated alternation of mechanical working and annealing reduces the size and flattens the platinum grains orienting them in the direction of hammering; a certain degree of solid diffusion of the platinum into the gold matrix occurs. (Scott y Bray 1980, 1994). The result is a coherent sinter with homogeneous hardness, silver or steel grey colour and mechanical and thermal conditions that permits cold working and soldering.

Aboriginal sintering was not that simple. On the basis of the technique described several variations were devised, including plating with platinum (platinum-clad alloys) and bi-metallic objects (Scott y Bray, 1980, 1994). We also have to consider the intentional addition of copper to the gold-platinum sinter that results in objects with a white pinkish colour; the effects of a prolonged heating of the mixture that renders it more homogeneous and in some cases possible absence or very limited mechanical work, as proposed by Bustamante et al. (2007). In this last case the procedure would have been as follows: 1) Making a mould, 2) Pre-selection of the material, 3) Filling the mould with a mixture of alluvial gold and platinum with a content of Pt not exceeding 50%, 4) Heating to the temperature of melting of gold and maintaining this temperature for a long time, 5) Cooling down and 6) Finishing by polishing and burnishing (Bustamante et al, 2007).

Estévez (2003) describes an extremely thin platinum plating (20 to 30 microns) that seems to have required the previous preparation of a foil of sinter 36.6% Pt in an alloy 36.5% Au, 21.9% Ag, 1.4% Cu and 3.6% Fe, that was afterwards adhered to a gold object. Scott y Bray (1980, 1994) point out the existence of a few objects that were made from grains of natural alloys or sinters of platinum – iron or platinum – copper – iron by means of simple mechanical work without intentional sintering. Finally there are processes of soldering of platinum sinters and platinum plating of gold objects about which there is no certainty.

Platinum technology is, as can be seen, quite complex. The major achievements were to overcome the difficulties imposed by the very high melting point of platinum and its hardness in order to exploit its chromatic qualities in a variety of manners. Thus it was possible to obtain platinum colour objects, objects that combine on the same side the colours of gold and platinum and objects with one golden side and the other platinum coloured. This technology was an ingenious adaptation to the natural conditions of the gold placers of the north coast that would otherwise be very difficult to use, as they were in fact for colonial miners for whom platinum or *platina* as it was named then, was highly inconvenient (Scott y Bray 1980, 1994).

The technology of gold working is equally varied and complex. To start with, very pure gold or gold naturally alloyed with silver in proportions of up to 27% Ag were frequently used (Estévez 1999). This is consistent with the known composition of placer gold from the region of Esmeraldas (Rivet et Arsandaux 1946). A range of gold – copper alloys, tombac, was also used, though less frequently. In some cases copper was added in low amounts (2 to 7%, Estévez, 2003), while in other objects it accounts for more than 90% of the object (Rovira 2004).

The two basic techniques, melting and hammering, were amply known and masterly employed, even though the second one is predominant. Perhaps the most impressive examples of hammering are the two funerary masks (Golden suns of Quito and Guayaquil), the largest one of which (3962-2-60) is 40 x 46 cs. and is made of a single sheet with a very uniform width, in spite of its intricate design with long rays decorated by repousse (Estévez 2002). The absence of stress fractures and the flexibility of the hammered sheets is consistent with annealing as the last stage in mechanical working;

the exceptions are tools and instruments (chisels, fishhooks, etc.) that show the temper given by hammering as the last stage in manufacture. The virtuosity of hammering was taken to unbelievable extremes, as was the case of the manufacture of very thin gold threads (Bouchard 1979). Those threads, as well as thicker wires were hammered and annealed leaving a cross section with facets; there are no evidences of the use of threaders or wire drawers or the use of abrasion as suggested by Zevallos (1956).

Even though there are no metallographic studies to back the idea, it is interesting to consider that hammered pieces were not always made from ingots or blocks previously melted. As a matter of fact it is probable that small or medium size objects may have been hammered directly from native gold grains that tend to appear near gold veins in relatively large sizes (<15 gs.). The other possibility is that alluvial gold may have been sinterised to make sheets omitting melting. In a process similar to that used to handle platinum, alluvial gold powder would have been heated and hammered until it was compacted; the resulting microstructure would show minute "blocks" that tend to separate leaving gaps and pores in the surface (Vera, com. pers.)

Hammering was used in conjunction with assemblage to make tri-dimensional objects from flat sheets (for example human figures), to plate objects made of other materials (i.e. snail shells) or to fix hanging elements (plaques, eyes, etc.) to masks, nose ornaments or other larger objects so as to add moving parts and produce metallic sounds with movement. In order to assemble mechanical means were used, such as flanges, sheet folding, hooks, rings and nails. Welding was the preferred metallurgical method for assemblage. A special type of assemblage was used to make needles; the pre-form, a hammered rod, was left with a wider part in the shape of fins near one of the ends. This end was then folded upon itself and the fins folded pressing the rod, thus forming the needle hole (Holm 1963).

In most tri-dimensional figures and masks there must have existed a wooden nucleus that rendered stability to the assemblage and that served to shape the sheets and hold them in place with nails. With flat figures, flanges, folding and rings were preferred for assemblage (Estévez in Pérez et al 1995). Frequently mechanical assemblages were combined with welding; this allowed the metal smiths to integrate miniature components of different metals.

The type of welding used in La Tolita – Tumaco took advantage of the differences in the fusion point of the god-copper alloys made with different proportions of the two metals (Scott y Doehne 1990). Some miniature objects, such as the skin appliques (face nails) were made of two melted parts that were then welded together; the primary components are made of alloys with fusion points higher than 1000 °C while the alloys used to weld melt between 870 and 940 °C; the smith could join the parts without melting them as long as he kept the welding temperature within a range of approximately 60 to 130 °C. (Scott y Doehne 1990).

Melting or casting of objects was done mainly by the lost wax method, with the exception of the small spheres made to be welded to larger objects. Those were made by the process of granulation; gold powder was heated on a ceramic mould with depressions the size of the spheres to be made (1 to 3 mms. in diameter), when they reached the fusion temperature the surface tension of molten gold would form the spheres in the depressions of the mould. There are no evidences of the use of moulds for other types of castings. The combination of casting and hammering is also frequent in the manufacture of objects of gold and tombac; objects made by casting could be retouched, modified or hardened, completely or partially, by hammering. The final stage in hammering was cutting, an operation that was done with metallic or stone chisels.

Finishing techniques include the already mentioned platinum platings and depletion gilding; applied to tombac objects and welding areas this treatment concealed the colour differences between components and welding material. There are no further evidences of the use of other types of gilding in this Group. Most La Tolita-Tumaco objects have an excellent surface polishing, no traces of working are visible and there are no burrs or parts of the casting conducts left on the finished objects.

For decorative purposes frequently repousse was used; in some cases it was done very deeply so that it formed huge convexities and concavities, as in the case of masks. This type of repousse implied working on both sides of the sheets with blunt instruments.

In La Tolita – Tumaco metal was combined with other materials, thus integrating several handicrafts. Valdez (1987) found a thin bundle of vegetable fibres, 10 cs. long, covered with gold dust. Inlays or danglers of precious or semi-precious stones also abound in objects of gold or gold-platinum. Among the stones used there are emeralds, apparently so frequent that they gave the region its present name. Turquoise, sodalite, serpentine, jadeite, agate, quartz and obsidian were used (Valdez, 1987). These stones were not cut in facets, as done presently, but rather rounded and polished; holes were drilled in a bi-conical shape.

Copper metallurgy has also attracted the attention of the researchers in the region (Reichlen 1942, Verneau et Rivet 1912, Bergsoe 1938, Rivet et Arsandaux 1946, Rovira 1990, Guinea 1998). Some of them provide composition analyses and metallographies of axes, chisels, needles, rings, nose ornaments and ear pendants made of copper; in a few cases it turns out to be almost pure copper (<99% Cu, in Reichlen 1942). Scott (1982) affirms that in La Tolita – Tumaco five different types of copper were used; native copper, arsenical copper, tin copper, brass and gold – copper alloy. The problem is that the contexts and dates of those objects are not known so that some of them can belong to Inca or colonial times.

Techniques used for copper manufacturing included also casting, hammering and a combination of both whenever necessary; cold working is dominant. There is a difference with respect to gold and tombac metallurgy in that for certain large copper objects, such as axes, it seems that bivalve moulds were used. This hypothesis is to be tested by the microscopic analyses of these objects. Rovira (1990) confirmed, by means of metallography, processes of cold working and annealing on alloys of arsenical copper, as well as the existence of previous casting stages. The most interesting statement of Rovira (1990), from his work with the objects of Atacames, is the presence of gilded copper; this author concludes that a process of electrochemical deposition was employed, something that is still to be confirmed.

A particularly striking feature of the La Tolita – Tumaco metallurgy is the use of lead. Rivet et Arsandaux (1946) report two small lead spheres from La Tolita. This should not be surprising since lead is present, as an impurity, in some copper and silver minerals and its fusion melting point is very low (328 °C), so that obtaining it is quite easy. The mechanical properties of leads do not make it appropriate for the manufacture of ornaments or instruments so that it may not have attracted the attention of metal smiths beyond the occasional manufacturing of simple small objects, like spheres. There is also a single reference (Rovira, 1990) of a silver object from that region; however it can be a colonial piece.

As previously stated, the absence of a classification of the objects of this Group into the Early and Late components makes understanding the evolution of metallurgical technology very difficult. The available evidences allow us to formulate a preliminary hypothesis in the following terms.

There are huge differences between the metallurgies of the Early and Late periods; in the Early Period the technology of platinum was developed and during the Late Period (after A.D. 700) it would have been largely or completely abandoned. During the Early Period the platinum technology was accompanied by gold working that would have continued in the Late Period, except that tombac would have been much more popular. The Late Period would have seen the introduction of copper; even though this metal was used in the Early Period it was then always alloyed with gold. During the Late Period copper was used unalloyed or alloyed with arsenic. In synthesis there seem to be some continuities, combined with changes in the emphasis of the use of metals and alloys that imply changes in the technologies of manufacture and finishing of objects.

Typology and classification

In the study of form and function in the La Tolita – Tumaco metallurgy, as shall be the case with all the other Great Regional Groups, we will take as our basis the collection of the Ministry of Culture of Ecuador (Quito). Whenever possible (as in the case of La

Tolita – Tumaco and Carchi – Nariño) we will use the collection of the *Museo del Oro de Colombia*. Only after the information from these sources has been exhausted we will recur to bibliographical sources to report complementary aspects and forms that are not represented in the collections. The following section is based upon the classification previously made by the scientists of the *Banco Central* because, as previously explained, it was not possible to check this typology against the objects.

In this sense we must recognise that the classificatory system of the collection of the Ministry of Culture is something that could certainly be improved and refined. A serious restriction that arises from the present classificatory system is the large number of objects categorised as "indefinable" that is objects whose function is unknown or has not been defined. In order to avoid altering the statistics of the objects of the Groups we have omitted from the study those objects, as well as the ones marked "various pieces" that usually group together under one inventory record various different objects and the "fragments"; none of these categories provides valuable information for our study. Due to these intentional omissions the total number of objects reported in this section is inferior to the numbers that were reported in previous sections. In the following table w summarise the categories of objects belonging to the La Tolita – Tumaco Regional Group in the collection of the Ministry of Culture, together with their frequency:

Function	Form - Representation	Frequency	Percentage
Needle	Simple – total	10	0.76
Wire	Simple	53	
	Double	2	
	With wide endings	2	
	With dangler	2	
	With adornment	7	
	Shaped as a hook	1	
	Twisted	9	
	Annular	11	
	Shaped as an "S"	2	
	In spiral shape	6	
	With bead	9	
	Wrapped	4	
	Total wires	108	8.24
Pin	Simple – total	7	0.53
Ring	Simple	22	
	Open	1	
	Hollow	1	
	With bead	1	
	Total rings	25	1.91
Fish-hook	Simple	12	
	Miniature	1	
	Total fish-hooks	13	0.99

Function	Form - Representation	Frequency	Percentage
Skin appliques	Simple	99	
	Hollow	1	
	Vegetable shaped	2	
	Shaped as a flower	1	
	Flat	1	
	Curve	8	
	In the shape of a double head	1	
	Total skin appliques	113	8.62
Shanks	Simple	43	
	Open with dangler	2	
	Rectangular	1	
	Hollow	1	
	Open	5	
	With bead	5	
	With dangler	1	
	With glass	1	
	Flat	2	
	From wire	1	
	Total shanks	62	4.73
Button	Simple – total	1	0.08
Lime container	Bi-conical	1	
	Globular	1	
	Total lime containers	2	0.15
Rattle	Simple	4	
	Zoomorphic, shaped as a head	2	
	Cylindrical	1	
	Total rattles	7	0.53
Helmet	Simple – total	1	0.08
Dangler	Simple	11	
	With dangler	2	
	Vegetable shaped	2	
	Conical	1	
	With adornment	1	
	With turquoise	1	
	With spiral adornment	1	
	Oval	10	
	Triangular	1	
	Bird shaped	1	
	Anthropomorphic, shaped as a head	1	
	Zoomorphic	1	
	With stone	1	
	Circular	3	
	Elliptical	1	
	Total danglers	38	2.90
Necklace	Simple – total	1	0.08
Bead necklace	Simple	9	
	Miniature	1	
	Globular	2	
	Semi-spherical	1	
	Total bead necklaces	13	0.99
Spoon	Simple	94	
	Shaped as a leaf	5	
	Total spoons	99	7.55

Great Regional Groups: La Tolita –Tumaco

Function	Form - Representation	Frequency	Percentage
Bead	Simple	37	
	Conical	2	
	Tubular	2	
	Annular	1	
	Oval	1	
	Various shapes	3	
	Vegetable shaped	2	
	Hollow cylindrical	5	
	Rhomboidal	1	
	Circular	13	
	Miniature	1	
	Semi-spherical	2	
	Globular	2	
	Total beads	71	5.42
Diadem	Simple	1	
	With danglers	1	
	Total diadems	2	0.15
Figure	Simple	4	
	Zoomorphic	2	
	Anthropomorphic	3	
	Total figures	9	0.69
	Simple	37	
	Concave	1	
	Vegetable shaped	1	
	With zoomorphic figure	2	
	Cylindrical	1	
	Narrow with ornament	1	
	Circular miniature	1	
	Rhomboidal	2	
	With wide endings	6	
	With anthropomorphic figure	1	
	Shaped as a snail	2	
	Annular	2	
	Cross shaped	1	
	Long	9	
	Circular	12	
	Rectangular	1	
	Vegetable shaped with dangler	2	
	With shank	5	
	Leaf shaped	7	
	Semi-spherical	4	
	Oval	2	
	Shaped as a spatula	1	
	Square	2	
	With holes	4	
	Anthropomorphic	1	
	Bird shaped with dangler	1	
	With stone	1	
	With adornment	2	
	Shaped as a horseshoe	1	
	Zoomorphic	3	
	Shaped as a feline	1	
	Total sheets	117	8.92

Function	Form - Representation	Frequency	Percentage
Mask	Simple	5	
	Anthropomorphic	1	
	Zoomorphic with dangler	1	
	Total masks	7	0.53
Miniature mask	Simple	2	
	Anthropozoomorphic	1	
	With adornment	1	
	With hanging adornment	1	
	Anthropomorphic	9	
	Anthropomorphic, shaped as a head	2	
	Zoomorphic	11	
	Bird shaped	1	
	Total miniature masks	28	2.14
Nose ornament	Simple	245	
	Hollow	18	
	With bead	23	
	Vegetable shaped	2	
	With adornment	8	
	Annular from wire	2	
	Narrow	2	
	Spiral shaped	1	
	Shaped as a double spiral	1	
	Triangular	1	
	Elliptical from wire	2	
	Annular	5	
	Zoomorphic double	1	
	Concave	1	
	From wire	8	
	With emerald	1	
	Circular with bead	3	
	Rattle	4	
	With holes	1	
	Elliptical with adornment	2	
	Circular hollow	4	
	Tubular	16	
	Crescent shaped hollow	1	
	Annular hollow	1	
	Elliptical hollow	7	
	Anthropomorphic, shaped as a head	1	
	Elliptical	6	
	Circular	7	
	Annular with lapis lazuli	1	
	Miniature	3	
	Long	1	
	Flat	9	
	Hollow with bead	1	
	With lateral extensions	1	
	In the shape of a crustacean	1	
	Elliptical with bead	1	
	With shank	3	
	Compound shape	1	
	Total nose ornaments	404	30.82

Great Regional Groups: La Tolita – Tumaco

Function	Form - Representation	Frequency	Percentage
Ear pendants	Simple	22	
	Cylindrical with rattle	3	
	Spiral shaped	9	
	From wire with adornment	1	
	Zoomorphic, shaped as a head	5	
	Elliptical with adornment	2	
	Anthropomorphic, shaped as a head	1	
	Tubular	10	
	With dangler	11	
	Tubular with bead	12	
	Double	2	
	Zoomorphic	3	
	Elliptical hollow	54	
	Elliptical	5	
	Annular with anthropomorphic dangler	1	
	Circular	1	
	With adornment	4	
	With lapis lazuli	2	
	Cylindrical	8	
	Anthropomorphic with dangler	1	
	Total ear pendants	157	11.98
Breastplate	Simple	1	
	Anthropomorphic	1	
	Total breastplates	2	0.15
Nipple cover	Simple – total	1	0.08
Tweezers	Simple	2	
	Circular	1	
	Total tweezers	3	0.23
Plaque	Simple – total	7	0.53
Vessel/bowl	Simple – total	2	0.15
Sun head shaped, anthropomorphic with zoomorphic figure		1	0.08
Total		1,311	100.00

The summary of form and function as shown in the previous table reveals some clear tendencies in the stylistic trends of the La Tolita - Tumaco Group that we shall discuss afterwards. At this point we can make use of the additional information provided by the collection of the *Museo del Oro de Colombia* from the southern Pacific coast, as summarised in the following table:

Function	Form - Representation	Frequency	Percentage
Sub-labial ornament	Shaped as a hook with dangler – total	1	0.49
Needle	Simple - total	2	0.97
Fish hook	Simple - total	21	10.19
Applique	Simple	2	
	Rhomboidal	7	
	Elliptical concave	2	
	Circular	5	
	Anthropomorphic	3	
	Shaped as a hook	1	
	Compound shape miniature	2	
	Total appliques	22	10.68

Function	Form - Representation	Frequency	Percentage
Skin applique	Shaped as a hook	3	
	Truncated cone	1	
	Compound shape	13	
	Shaped as a flower	1	
	Spiral	1	
	Rhomboidal	1	
	With circular ending	10	
	Total skin appliques	30	14.56
Textile applique	Simple	2	
	Circular concave	1	
	Compound shape	5	
	Square	5	
	Compound shape with spiral ornament	1	
	Circular	6	
	Elliptical	1	
	Crescent shaped, in the form of a raft	5	
	Total textile appliques	26	12.62
Rattle	Circular - total	1	0.49
Dangler	Annular	2	
	Compound shape	1	
	Zoomorphic, shaped as a fish	1	
	Square with ornaments	1	
	Elliptical concave	1	
	Total danglers	6	2.91
Ear pendant	Circular	1	
	Rectangular with zoomorphic figure	2	
	Circular anthropomorphic	2	
	Total ear pendants	5	2.43
Bead necklace	Globular - total	1	0.49
Bead	Simple	1	
	Cylindrical	1	
	Zoomorphic schematic	1	
	Total beads	3	1.46
Diadem	Anthropozoomorphic - total	1	0.49
Figure	Zoomorphic, shaped as a fish	1	
	Anthropomorphic	3	
	Total figures	4	1.94
Sheet	Circular	1	
	Triangular	1	
	Total sheets	2	0.97
Mask	Anthropomorphic	1	
	Anthropomorphic miniature	2	
	Total masks	3	1.46
Nose ornament	Annular	33	
	Crescent shaped	5	
	With ascending extensions	1	
	Crescent shaped with horizontal extensions	2	
	Circular	2	
	Shaped as "n"	1	
	Annular with ornament	1	
	Total nose ornaments	45	21.84

Function	Form - Representation	Frequency	Percentage
Ear pendant	Shaped as a hook with dangler	6	
	Annular	2	
	Elliptical	6	
	Annular with dangler	2	
	Shaped as a hook with ornament	2	
	Elliptical with dangler	2	
	Crescent shaped hollow	1	
	Annular with ornament	1	
	Total ear pendants	22	10.68
Breastplate	Anthropozoomorphic with spiral ornament - total	1	0.49
Tweezers	Semi-globular - total	1	0.49
Plaque	Circular miniature	1	
	Rectangular	1	
	Circular	1	
	Total plaques	3	1.46
Plaque pendant	Circular concave	1	
	Rectangular	1	
	Trapezoidal	1	
	Compound shape	1	
	Total plaque pendants	4	1.94
Vessel/ bowl	Compound shape - total	1	0.49
Ear pendant lid, square with zoomorphic figure - total		1	0.49
Total		206	100.00

Except for minor variations, due probably to differences in the policies of acquisition of the two museums and the application of slightly different classificatory principles, in both collections we can appreciate very similar trends. There is a vast majority of adornments for the head and face; in both collections the skin appliques ("facial nails") that were used mainly, though not exclusively on the face, the sub-labial adornments, helmets, diadems, masks, nose ornaments and ear pendants represent respectively 52.18% and 63.12% of the samples. Furthermore, these categories of objects, especially skin appliques, nose ornaments and ear pendants are more varied and better decorated in spite of being quite small. The exceptions, with regard to size, are the diadems known as the golden suns, which are large and very complex objects.

The types of objects used on the trunk such as necklaces and beads, breastplates and nipple covers account for 9.62% (*Ministerio de Cultura*) and 5.35% (*Museo del Oro*) respectively and are restricted with regard to shapes and decoration. Not even large objects like breastplates are too elaborate. The group of objects worn on the hands like rings and shanks account for 6.64% (*Ministerio de Cultura*) and are absent in the other collection. There are finally other objects that are not worn on any specific part of the body (textile appliques, rattles, buttons and plaque pendants) that are quite scarce. In general terms the Group is characterised by a very marked trend to adorn the face and head with a much lower emphasis on the upper trunk and hands and a virtual absence of adornments for the lower trunk and legs.

Other metallic objects belong to the group of tools and instruments; in this group there are needles, pins, fish hooks, lime containers, spoons, tweezers and bowls that account for 10.36% (*Ministerio de Cultura*) and 12.14% (*Museo del Oro*). Sheets and wires that are abundant in both collections are problematic; they might include objects that were erroneously classified, unfinished objects, excess material or recycled raw material from workshops (Bergsoe 1938), objects that were intentionally deformed or broken ("killed") or parts of larger objects. This uncertainty prevents us from establishing and analysing their frequency.

The last interesting and scarce category in both collections are figures, both anthropomorphic and zoomorphic (fishes, snails). These were made by mechanical assemblage of sheets over wooden cores; some have survived almost complete and others only in parts like arms, ears, noses, etc. In the same category we have the miniature masks that, due to their size, were not made to be used by persons. They must have been parts of complete metal, clay-metal or wood-metal figures. Those objects boast an elaborate manufacturing and decoration; they repeat, at a smaller scale, the emphasis of the head and face as subjects of ornament.

In the collection of the Ministry of Culture there are also axes and blowing pipes (nozzles) belonging to the La Tolita – Tumaco Group that are not classified as such in the inventory. Apart from the objects described in the collections, for this region researchers mention axes, chisels and drills (Bergsoe 1937-38/1982), palettes (Guinea 1998), bracelets, anklets, hooks and small copper anvils (Valdez 1987).

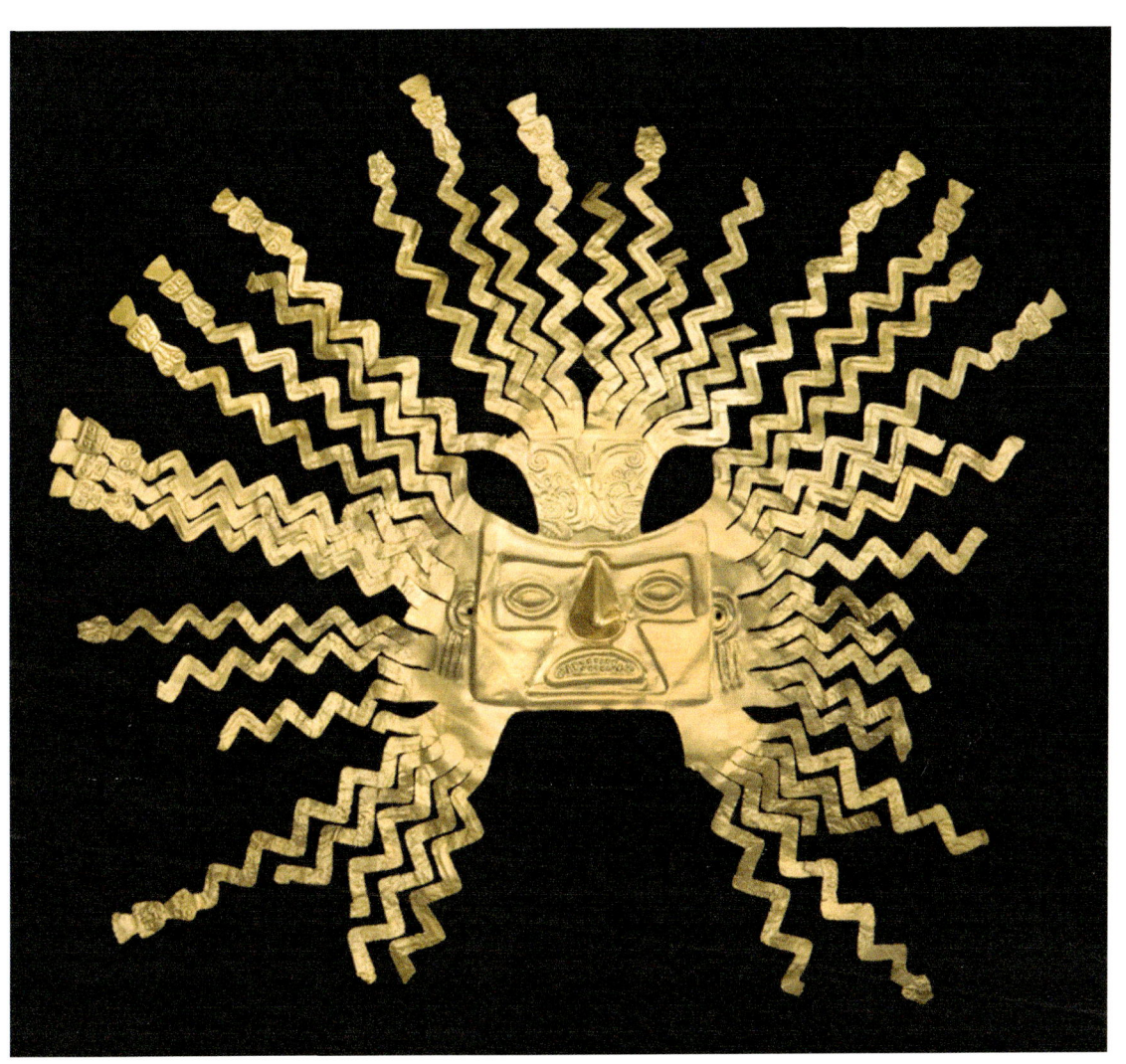

FIGURE 8 LA TOLITA – TUMACO GOLD ANTHROPOMORPHIC MASK WITH EXTENSIONS IMITATING THE RAYS OF THE SUN: 40 X 60 X 0.3 CMS.

Figure 9 La Tolita – Tumaco gold and platinum zoomorphic mask: 7.5 x 5.5 x 4 cms.

GREAT REGIONAL GROUPS: LA TOLITA –TUMACO 53

FIGURE 10 LA TOLITA – TUMACO GOLD AND PLATINUM WITH SODALITE INLAYS
ANTHROPOMORPHIC MASK: 9.6 X 9 X 5.6 CMS.

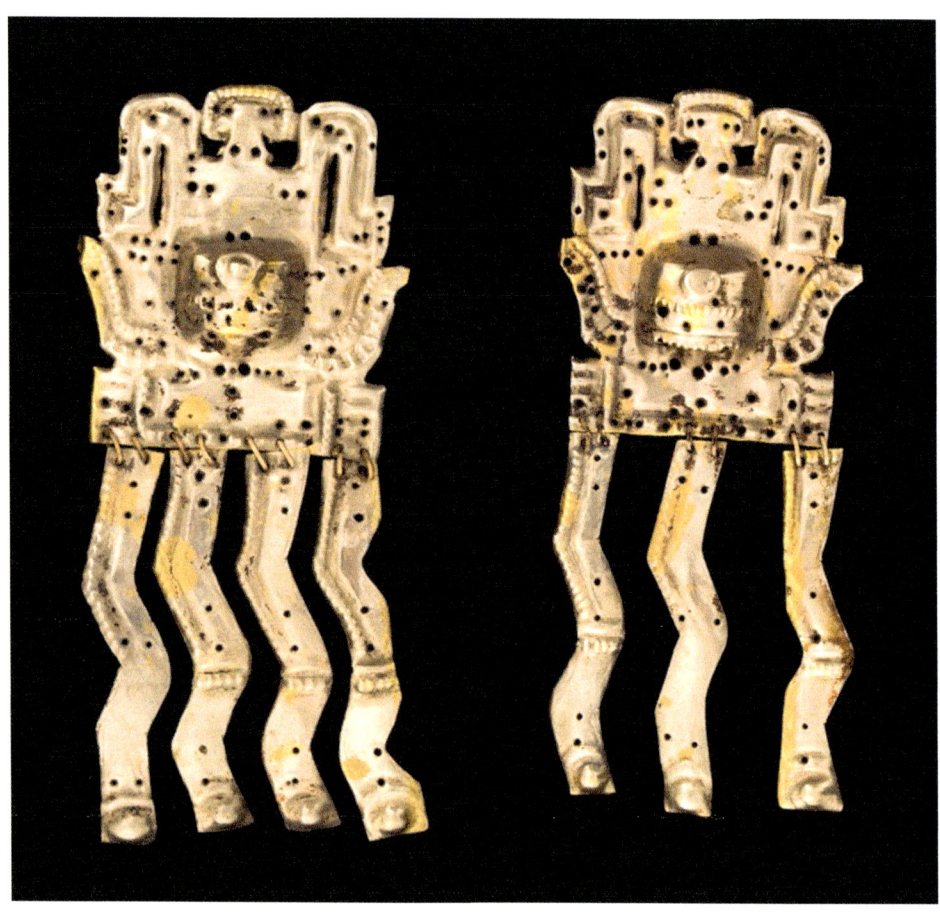

Figures 11 and 12   La Tolita – Tumaco gold ear pendants: 16 x 4.6 x 1.2 and 16.2 x 4 x 1.2 cms.

Great Regional Groups: La Tolita –Tumaco

Figure 13 La Tolita – Tumaco gold and platinum zoomorphic mask, two components: 4.7 x 8.1 x 0.3 and 5.3 x 7.7 x 2.6 cms.

Figure 14 La Tolita – Tumaco gold and platinum anthropomorphic mask: 19.8 x 19.8 x 9.6 cms.

Great Regional Groups: La Tolita –Tumaco 57

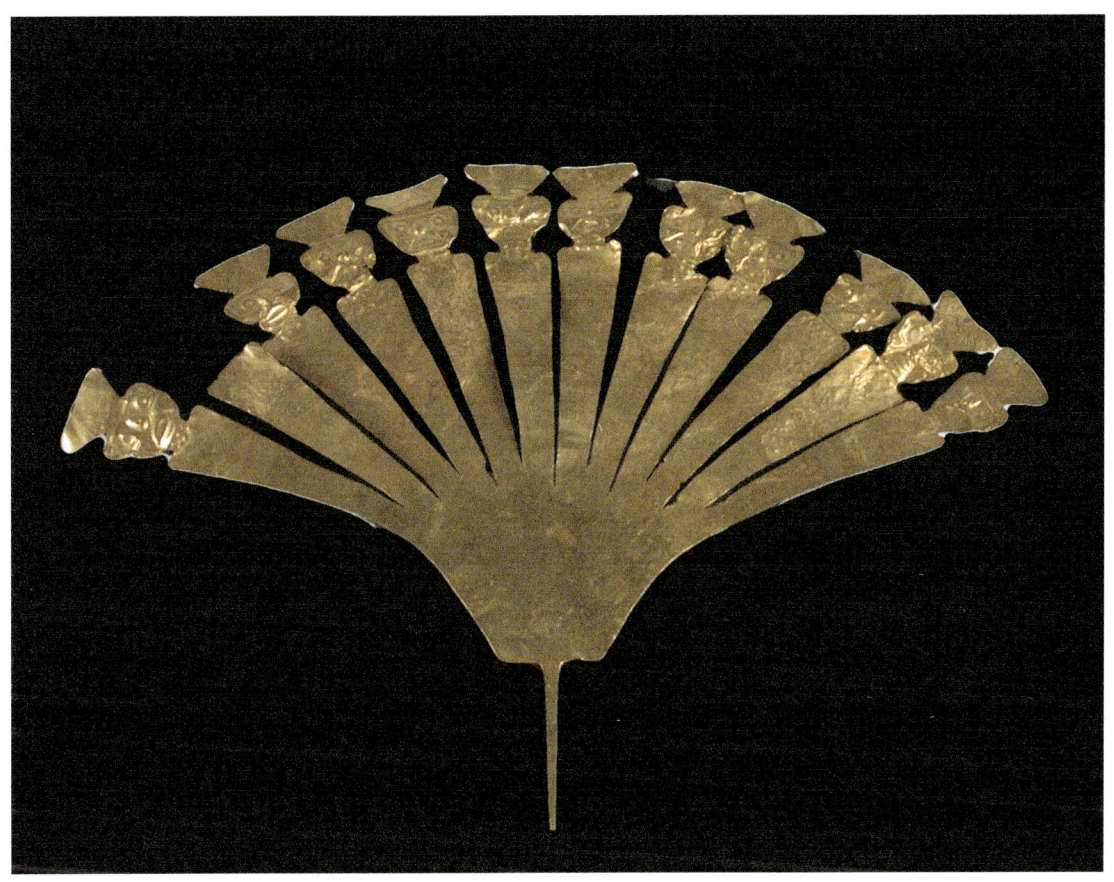

Figure 15 La Tolita – Tumaco gold crest for diadem: 22.3 x 32 x 0.2 cms.

58   Metallurgy in Ancient Ecuador

Figure 16 La Tolita – Tumaco gold necklace: 1.9 x 78 x 0.1 cms.

Great Regional Groups: La Tolita –Tumaco   59

Figure 17 La Tolita – Tumaco gold zoomorphic figure: 4.4 x 4.2 x 21.7 cms.

Figure 18   La Tolita – Tumaco copper axe: 11 x 7 x 2.5 cms.

Chapter 6

Great Regional Groups: Jama – Coaque

The Jama – Coaque Regional Group is, unlike the preceding Group, relatively unknown. The archaeological expeditions in this area have not recovered object in situ that may provide information about the chrono-stratigraphic contexts, while the Jama – Coaque collections have failed to raise much interest among archaeo-metallurgists. The collection of the Ministry of Culture is quite small.

Geographical distribution

The following table summarises the information concerning the provenance of the objects catalogued as Jama – Coaque in the collection of the Ministry of Culture (Quito):

Province	Canton	Number of objects	Percentage of total
Manabí	Jama	24	13.8
	Pedernales	2	1.1
	San Isidro	143	82.2
	Sucre	5	2.9
	Total	174	100

The small size of the sample is a problem for statistical purposes. To start with it is very unlikely that the Jama – Coaque metallurgy would have been found only in four sites of the Manabí province and that over 82% of the objects come from only one canton. Evidently there is a problem besides that of the size of the collection and it has to do with the declaration of provenance. The registration of relatively large numbers of objects, declared as found in San Isidro, completely distorts the map of distribution.

The following map, drawn on the basis of the available information, displays the distribution of objects from the Jama – Coaque Regional Group (green circles) with relation to the provenances of all the other objects of the collection of the Ministry of Culture (Quito), (red circles):

The deficiency of the information prevents us from inferring a real pattern of distribution; the only thing that can be deducted at a mere indicative level is that the objects from this group come from sites located in or near the seashore in the north of the province of Manabí (Jama, Pedernales and San Isidro). There is only one site (Sucre) with a very small number of objects (2.9%) located in the same province but further inland than the other three sites. Our provenance information for this Regional

Metallurgy in Ancient Ecuador

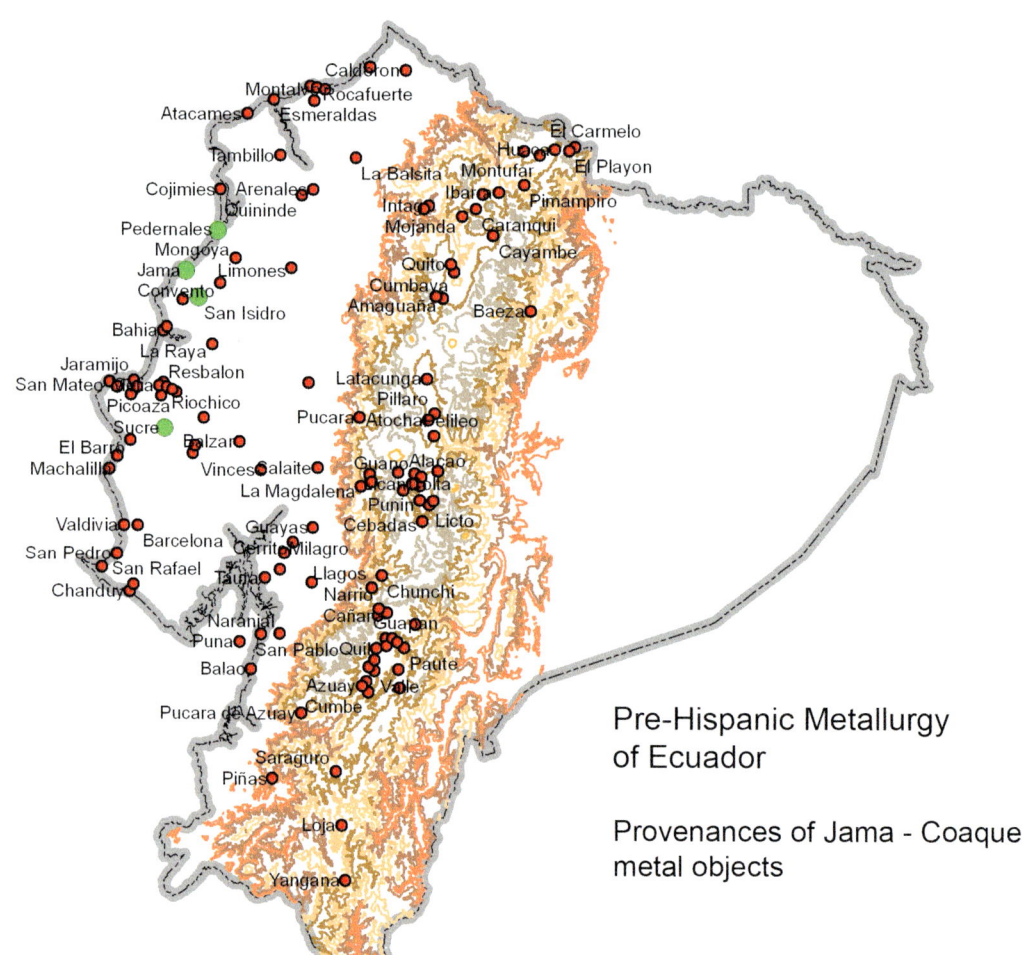

Pre-Hispanic Metallurgy of Ecuador

Provenances of Jama - Coaque metal objects

Group is still precarious so that it would be convenient to research about finds in the nearby villages (Coaque, Bahía, Río Mariano, San Vicente, etc.).

Chronology

There are no absolute dates associated to metal objects of the Jama – Coaque Group, either from excavations or collections. There are many C14 dates from the Jama valley covering a time span from 1700 B.C. to A.D. 1300 (Radiocarbon Database, 2006) but apparently none is associated to metal objects. The chronological estimate for Jama – Coaque used at present by the *Museo Nacional del Ecuador* that goes from 400 B.C. to A.D. 1533 (Pérez et al 1995) is excessively long. It does not seem probable that the Jama – Coaque metallurgy would have lasted for such a long time. Even though its initial phases may have been more or less contemporary with those of La Tolita – Tumaco or slightly later, there are no evidences of the development of a late phase, since its technological and iconographic characteristics are quite homogeneous. It seems much more plausible that this metallurgy was replaced in the area by the Manteño - Huancavilca metallurgy during the 7th century, at the latest. Our estimate, that for the time being is just that, would be in the range of 400 B.C. to A.D. 600.

Technology

Very little is known about the Jama – Coaque technology in comparison to that of the La Tolita – Tumaco group. To start with the Group seems to be quite homogeneous from the technological point of view. Silver and platinum are completely absent. Almost all objects are made of gold or gold alloys with the exception of a copper pendant found in San Isidro. Barrandon, Valdez y Estévez (2002) provide composition data for three Jama – Coaque objects. Two of the samples are alluvial silvery gold with small amounts of copper (3.7 and 4.3%); trace elements include platinum and platinoids, something to be expected in the alluvial placers of this area. The other sample is a tombac with 52.8% Cu and 42.8% Au (the remaining is silver). This scarce data, together with the observation of the surface colour of the objects, suggest that in Jama – Coaque gold and copper-gold alloys were used, including tombacs with more than 50% copper.

Most objects were made by hammering; casting was employed only for the manufacture of small components. The combination of both basic techniques is not frequent. To assemble they used flanges and hooks, there is also evidence of small welding points, but we do not know what type of welding was used. By means of assemblage they made tri-dimensional figures that must have had wooden nuclei; the most frequent type of assemblage is related to danglers. To decorate some objects they used granulation to adhere small spheres.

Sheets and wires were made by hammering, the later were figured to form spiral nose ornaments. For decoration the most frequent technique was repousse, involving working both sides; complex motifs were embossed with mastery and precision. Thin sheets were used to plate or cover organic materials, some of which have decayed. Objects show a good degree of polishing; there are no burrs or traces of hammering. Semi-precious stones were integrated to gold adornments as danglers or inlays; those included turquoise, serpentine, jadeite and quartz.

Typology and classification

The repertoire of the Jama – Coaque objects analysed in this section is, most surely, just a fraction of those produced and used in pre-Hispanic times. An additional difficulty is that we are at present unable to understand and interpret the use of many objects that we cannot locate within any functional category. Nevertheless the research results open an insight into what must have been a rich and varied tradition. In the following table we summarise the forms and functions of the Jama – Coaque objects, together with their frequency in the collection of the Ministry of Culture (table opposite).

In most cases the analysis of the Jama – Coaque Group has been carried out simultaneously with that of the La Tolita – Tumaco Group (Liu 1992, Pérez et al 1995). Such approach is valid as long as it is not limited to point out the similarities but also the differences. From this point of view we can start by stating that the pattern observed in La Tolita – Tumaco that emphasises on the decoration of the face and head is even stronger in Jama – Coaque. Objects made to be worn in this part of the body (nose ornaments, ear pendants and plaques) add up to 85.17% of the simple, a proportion that absorbs almost entirely the repertoire.

O the other hand, the repertoire in this Group is even more restricted than in the preceding one; there are no skin applications (face nails) or rattles and fish hooks, etc. The only other category that has some quantitative importance is made up of hand adornments (rings and shanks) with 9.03%. Another important trait is the disappearance of large objects; except for a large zoomorphic breastplate, the objects are small or medium (>10 cms.). The Jama – Coaque metallurgy seems to be exclusively oriented towards facial ornament with nose ornaments and ear pendants and only secondarily to the adornment of the hands.

In the collection of the Ministry of Culture there are also shell covers and a staff that are not registered as such in the inventory. Some other objects such as helmets, bracelets, anklets and bowls are mentioned (Pérez et al 1995). Liu (1992) also mentions diadems, masks, appliques and bracelets represented in clay figurines, but warns that they might be metal or shell ornaments, so it is not safe to add them to our repertoire until we can see the real objects.

Function	Form - Representation	Frequency	Percentage
Wire	Circular with bead - total	3	1.94
Ring	Simple - total	1	0.64
Shank	Tubular	12	
	From wire	1	
	Total shanks	13	8.39
Lime container	Globular - total	1	0.64
Dangler	Simple - total	1	0.64
Bead necklace	Simple - total	1	0.64
Bead	Semi-spherical - total	1	0.64
Nose ornament	Simple	13	
	Hollow	5	
	Semi-circular hollow	1	
	Elliptical hollow	22	
	With wide endings	1	
	With adornment	6	
	Tubular	16	
	Circular hollow	3	
	Circular	4	
	Annular hollow	4	
	Truncated cone hollow	2	
	With bead	2	
	Tubular with bead	1	
	Elliptical concave	1	
	With turquoise	1	
	Zoomorphic	1	
	Trapezoidal hollow	2	
	Total nose ornaments	85	54.84
Ear pendant	Simple	4	
	With dangler	2	
	Elliptical hollow	10	
	Tubular	18	
	Annular	2	
	Elliptical with dangler	6	
	Tubular with bead	2	
	Total ear pendants	44	28.39
Breastplate	Zoomorphic – total	1	0.64
Tweezers	Elliptical concave – total	1	0.64
Plate	Zoomorphic - total	3	1.94
Total		155	100.00

FIGURE 20 JAMA – COAQUE GOLD PENDANTS: 9 x 7 x 6; 13 x 7.5 x 4 AND 12.8 x 7.9 x 4.4 CMS.

GREAT REGIONAL GROUPS: JAMA – COAQUE

FIGURE 21 JAMA – COAQUE GOLD BOWL: 3.4 X 12 CMS.

Figures 22 and 23 Jama – Coaque gold ear pendants: 10.9 x 3.9 x 3.7 and 10.6 x 4 x 3.6 cms.

Great Regional Groups: Jama – Coaque

Figure 24 Jama – Coaque gold breastplate with zoomorphic figure:
17.8 x 18 x 3.1 cms.

Chapter 7

Great Regional Groups: Bahia

The Bahia Regional Group is represented in the collection of the Ministry of Culture by a limited number of objects, some of which have no precise provenance information. The scarcity of chronological information, the geographical distribution and the variety of metals and alloys used pose several questions that we shall discuss in the relevant sections.

Geographic distribution

In the following table we can see the quantitative and percentage distribution of the metallic objects classified as belonging to the Bahia Regional Group in the collection of the Ministry of Culture (Quito):

Province	Canton	Number of objects	Percentage of total
Unknown	Unknown	4	1.3
Esmeraldas	Atacames	1	0.3
	Calderón	5	1.7
	La Balsita	5	1.7
	La Tolita	1	0.3
	Los Arenales	1	0.3
Guayas	Barcelona	1	0.3
Manabi	Manabi undefined	95	32.2
	Jaramijo	4	1.3
	Las Chacras	13	4.4
	Mejía	3	1.0
	Pedernales	2	0.7
	Picoaza	1	0.3
	Resbalón	3	1.0
	Riochico	7	2.4
	San Isidro	66	22.4
	San Vicente	4	1.3
	Salaite	79	26.8
	Total	295	100

The map displays the distribution of the Bahia objects (green dots) with relation to the distribution of the objects of all the other Regional Groups (red dots):

Great Regional Groups: Bahia

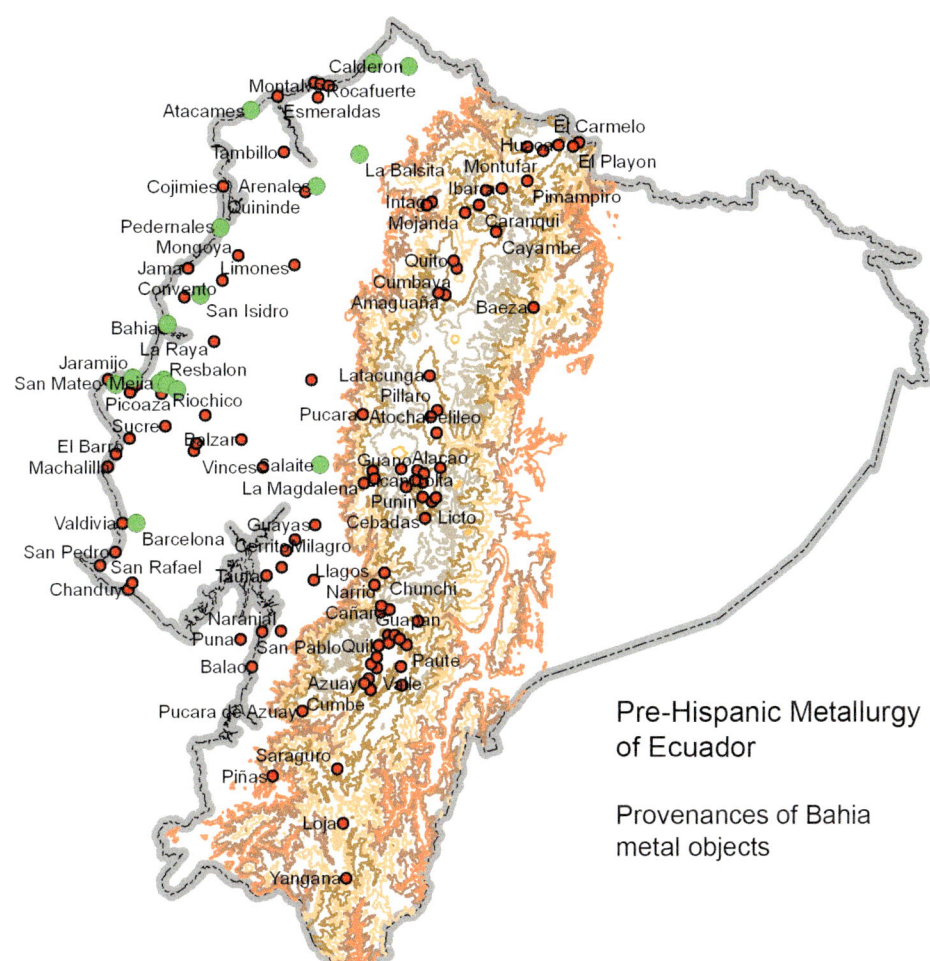

Pre-Hispanic Metallurgy of Ecuador

Provenances of Bahia metal objects

One third of Bahia objects have no provenance or an imprecise provenance, thus considerably reducing the sample with which we can attempt to define a spatial pattern of distribution. For the remaining pieces (196) this distribution pattern is quite odd. There are only two significant concentration foci (San Isidro and Salaite), both in the central Manabi seashore. Other objects appear in small groups, from Esmeraldas to the north of Guayas. It is difficult to assume any conclusions from this information that might correspond in part to imprecise provenance information as declared by sellers. The general trend seems to pint to a large distribution area, but this would have to be confirmed on the basis of a more solid statistical basis.

Chronology

Bahía holds a sui generis position in the Coast, both from the points of view of technology and form-function; from one side it is clear that it shares some of the most evident traits of the La Tolita – Tumaco and Jama – Coaque Groups and that, from there it developed new characteristics, such as the use of silver. On the other side it is also true that those new traits are much more developed in the Milagro – Quevedo and Manteño – Huancavilca Groups.

In other words the Bahia Group seems to constitute a transition between the two Early and the two Late Groups of the Coast. Should we accept this hypothesis the necessary consequence is that its chronology should be directly related to that of the Early Groups as well as to that of the Late Groups. The terminal dates of the Early Period of La Tolita – Tumaco Group are in the region of A.D. 200 and those of Jama – Coaque around A.D. 600, while the initial dates for Milagro – Quevedo are around A.D. 100 and those of Manteño – Huancavilca are around A.D. 600. It is possible that the Bahia metallurgy would have developed at the same time as the final manifestations of the Early Groups, perhaps around the first century before the Common Era (100 B.C.). We have, however, a single very late C14 date associated to an object in the collection of the Ministry of Culture:

Date	Canton	Laboratory Number	Reference
A.D. 1520 ± 40	M-Undefined	Beta 237173	Museo del Oro 2007

This preliminary result by itself is not enough to propose a radical reformulation of Bahia chronology; for the time being we can only state that we should leave open the possibility that this particular metallurgical tradition might have survived up to the time of the European conquest. Thus we believe that the estimate given by the Banco Central that situates this tradition between 600 B.C. and A.D. 600 (Pérez et al 1995) is excessive with regard to the initial date and falls short with relation to the final date that could extend considerably.

Technology

The technology of Bahia metallurgy has been studied only recently. Barrandon, Valdez y Estévez (2002) report XRF and NAA analyses of four objects from the collection of the Ministry of Culture. The XRF results (surface composition) show that three objects are silvery gold and copper alloys in proportions that range from 56 to 88% Au and 5.2 to 35.3% Cu with silver contents between 5.4 to 34.6%. The other object is made from silvery gold without added copper (88.6% Au, 9.3% Ag). The NAA results (bulk composition) agree in general with the previous ones, except that they show higher copper contents. This might indicate processes of surface depletion gilding or, alternatively, the loss of copper due to natural corrosion in the burial environment. Anyhow, the available data, together with the observation of the objects of the collection of the Ministry of Culture allow us to conclude that in Bahia metallurgy alluvial gold and copper – gold alloys (tombac) of different composition were used. The other metals used less frequently were copper and silver, the later one possibly alloyed with copper.

The predominant technique is hammering, that was handled with mastery in most cases, even in the case of difficult to work alloys such as silver – copper. Casting is scarce and was mainly used for small objects. Using hammered sheets they managed to make complex tri-dimensional objects, such as the silver figure shaped as a raft (98D-1-1); for this purpose they used mechanical and metallurgical assemblage techniques. The mechanical techniques included long wires with which they tied parts together. The welding of gold, silver and copper is also frequent; no metallographic studies have been done that might help us understand the type of techniques used in each case, except for the welding of small spheres that seems to correspond to the well-known technique of granulation. There is another assemblage technique that is not strictly mechanical; the piecing together of small cylindrical beads of silver into a large objects by means of cotton threads (Object 70-90-1).

In the collection of the Ministry of Culture there is, at least, one object made of gold plated copper; it may have been plated by fusion gilding or sheet plating. In the first case gold powder (with small amounts of copper to reduce the melting point) would have been heated until melted over the surface of the copper giving rise to some degree of solid inter-diffusion. In the second case a thin sheet of gold would have been placed over the surface and adhered by means of heat and hammering (Scott 1985). Bahia metal smiths must have mastered one, or perhaps both techniques; we need to go further into this problem.

Zevallos (1965) reports a copper mask with pottery ear pendants. The mask was made by hammering and showed a deficient technique evident in many fractures around the nose area. This would be an exception in the general high level of technology in Bahia.

For decorative purposes they used repousse, working on both sides of the sheets, generally over small areas and shaping intricate designs. Another remarkable trait of decoration is the use of semi-precious stones; those were inlayed or hanged with metal hooks and wires. Stones used include turquoise, jadeite and sodalite. The objects, in general, display good polishing there are no visible marks of hammering in the surface.

Typology and classification

With the advent of Bahia, metallurgy in the Coast acquires a new dimension. The flourishing and evolution of the La Tolita – Tumaco metallurgy, weaker in Jama – Coaque, regains in Bahia its former strength; new metals and alloys are used and there is a wider spatial distribution. With regard to forms and functions this new impulse can be summarised in the following table:

Function	Form - Representation	Frequency	Percentage
Sub-labial ornament	Spool shaped	5	
	Hollow	2	
	Shaped as half spool	1	
	With spheres	1	
	Total sub-labial ornaments	9	4.07
Wire	Circular with stone	2	
	With spheres	2	
	Annular	3	
	Annular with bead	1	
	Total wires	8	3.62
Ring	Simple - total	1	0.45
Skin application	Simple - total	8	3.62
Shank	Simple	7	
	With turquoise	2	
	With nail	1	
	With bead	2	
	With stone	1	
	Total shanks	13	5.88
Button	Semi-spherical - total	1	0.45
Chisel	Simple - total	1	0.45
Dangler	Simple - total	2	0.91
Ear pendant	Bi-conical - total	2	0.91
Bead necklace	Simple	1	
	With stone	1	
	With bead	1	
	Spiral	1	
	Total bead necklaces	4	1.81
Bead	Simple	2	
	Circular	4	
	With turquoise	1	
	Star shaped	1	
	Total beads	8	3.62
Disc	With holes - total	2	0.91
Spatula	Simple - total	1	0.45
Figure shaped as a raft with anthropomorphic figures		1	0.45
Figure	Anthropomorphic - total	1	0.45

Function	Form - Representation	Frequency	Percentage
Sheet	Simple	4	
	Cylindrical wrapped	2	
	Circular	6	
	Semi-spherical	1	
	Total sheets	13	5.88
Nose ornament	Simple	31	
	Elliptical hollow	7	
	With dangler	1	
	Circular double	1	
	Hollow	3	
	Careened	1	
	With bead	7	
	Spiral	3	
	Double spiral	2	
	Circular	12	
	Annular with bead	1	
	Annular	1	
	Circular hollow	2	
	Zoomorphic	4	
	Tubular	1	
	Anthropozoomorphic	2	
	Rattle	1	
	Phytomorphic	2	
	Semi-circular	1	
	Total nose ornaments	83	37.56
Ear pendant	Simple	3	
	Tubular	21	
	Spool shaped	3	
	Truncated cone	3	
	Cylindrical with shanks	1	
	Elliptical	2	
	With shanks	3	
	With beads	2	
	Cylindrical	2	
	Circular	2	
	Zoomorphic	2	
	Spiral	4	
	Tubular with shanks	1	
	Semi-spherical	1	
	Total ear pendants	50	22.62
Breastplate	Simple	1	
	Circular	4	
	Bevelled	1	
	Total breastplates	6	2.72
Tweezers	Simple	2	
	Shaped as a J	1	
	Circular with ornament	1	
	Total tweezers	4	1.81
Plaque	Rectangular - total	2	0.91
Container/bowl	Simple	1	
	With holes	1	
	Total bowls	2	0.91
Total		221	100.00

In spite of the evident change of technological trend in the Bahia metallurgy with respect to the preceding groups of the Coast, we can see that the decorative pattern that emphasizes on the head and face is kept; the objects designed to be worn in this part of the body (sub-labial ornaments, skin applications, ear pendants and nose ornaments) account for 68.78% of the sample. Nose ornaments and ear pendants are the most elaborate and varied objects; these two categories have, respectively 19 and 14 varieties. The rest of the body ornaments (rings, shanks, necklaces, beads and breastplates) are relatively scarce and simple.

The remaining objects belong mostly to problematic categories (wires and sheets) and account for 9.5% of the simple. There are also a few tools and instruments (chisels, spatulas and bowls) and two figures, one of them anthropomorphic and the other one an exceptional silver object representing a raft made of logs tied together with a principal person standing in the middle accompanied by two oarsmen and a helmsman.

Even though in the inventory it is not classified as such, in the collection of the Ministry of Culture there is a chest guard made of tubular silver beads tied together with cotton threads; there are also a few diadems. Zevallos (1965/66) reports a Bahia silver mask; Pérez et al (1995) mention the existence of bracelets, anklets and lime containers.

FIGURE 26 BAHIA SILVER VOTIVE FIGURE SHAPED AS A RAFT: 6.8 x 10 x 19 CMS.

Figure 27 Bahia gold snail cover: 9.5 x 10.2 x 21.2 cms.

Great Regional Groups: Bahia 79

Figure 28 Bahia silver chest guard: 8.7 x 17.4 x 0.3 cms.

Figure 29 Bahia gold pair of ear pendants: 12.8 x 3.4 x 3.3 and 12.6 x 3.7 x 3 cms.

Chapter 8

Great Regional Groups: Milagro – Quevedo

The Milagro – Quevedo Group, represented in the collection of the Ministry of Culture by a medium sized simple, has a meaningful volume of context information provided by archaeological research carried out in the Guayas – Daule basin from long ago.

Geographic Distribution

In the following table there is a summary of the quantitative and percentage distribution of the metal objects classified as part of the Milagro - Quevedo Regional Group in the collection of the Ministry of Culture (Quito):

Province	Canton	Number of objects	Percentage of total
Unknown	Unknown	69	18.5
Azuay	Sigsig	1	0.3
Cañar	Biblian	2	0.5
	Cañar	2	0.5
Carchi	El Angel	2	0.5
	Carchi undefined	1	0.3
Esmeraldas	Esmeraldas	15	4.0
	La Balsita	38	10.2
	La Tolita	1	0.3
	Tambillo	8	2.7
Guayas	Balao	2	0.5
	Balzar	89	23.9
	Boliche	2	0.5
	El Triunfo	1	0.3
	Guayas	18	4.8
	Guayas undefined	8	2.7
	Milagro	2	0.5
	Naranjal	6	1.6
	Santa Elena	12	3.2
	Taura	5	1.3
Manabí	Manabí undefined	16	4.3
	Mejía	14	3.8
	San Isidro	3	0.8
Los Ríos	Los Ríos undefined	44	11.8
	San Camilo	4	1.1
	Vinces	5	1.3
Tungurahua	Atocha	2	0.5
	Total	372	100

A considerable part of the collection (almost 38%) has no provenance information or has only very imprecise information referred only to the general area of a province (Manabí, Los Ríos, Guayas and Carchi). Particularly surprising is the existence of finds outside the historically known territory of the Milagro – Quevedo culture, especially the finds of the Sierra (provinces of Tungurahua, Cañar and Carchi). Even though such finds may derive from pre-Hispanic exchange, the doubt remains as to the possible misclassification of the objects or wrong provenance information; distortions in distribution information may arise from such mistakes. The Sierra – Coast interactions are well documented for the late period of Ecuadorian prehistory (Marcos 1995) but those processes seem to have occurred mainly in the central and southern parts of the country; in this sense we have to be cautious about this extra-territorial provenances.

In the following map the distribution of Milagro – Quevedo metallic objects is displayed (green dots) in comparison to the distribution of objects from all the other Groups (red dots) as recorded in the collection of the Ministry of Culture (Quito):

If we are to believe what seems to be the most trustable information, that covers about 60% of the sample, leaving aside the doubtful finds of the Sierra, we find that there is an exceptionally widespread distribution of Milagro – Quevedo objects. This distribution covers almost the entire Coast, from Esmeraldas to the south of Guayas with the sole exception of the province of El Oro. Within this extensive zone there are two areas of concentration; one of them, as was expected, is the basin of the Guayas River and the adjacent seashore (Balao, Balzar, Boliche, El Triunfo, Guayas, Vinces). The other concentration focus is located to the north, in the province of Esmeraldas (La Balsita, La Tolita, Tambillo). In the rest of the coastal area there are disseminated finds.

Taking into consideration what has traditionally been considered as the territory of the Milagro – Quevedo culture, this distribution would not be coherent. However, beyond the errors induced by wrong provenance declarations, these finds might be related to two different socio-political phenomena that would account for the existence of metal objects outside the basin of the Guayas-Daule. The first one would have to do with exchange processes as discussed by Marcos (1995) and the second one could be related to the expansion and contraction of the territory of the Milagro – Quevedo culture along the centuries. It is possible that the producers of this metallurgy could have occupied, before the XVI century, areas beyond the Guayas – Daule basin and that is why those objects would appear in those sites. Both hypotheses would need to be tested by field research.

Chronology

The only available dates for Milagro – Quevedo metallurgy come from two different digs, the first one from the La Compañía site (Meggers 1966 and Radiocarbon Database

Great Regional Groups: Milagro – Quevedo

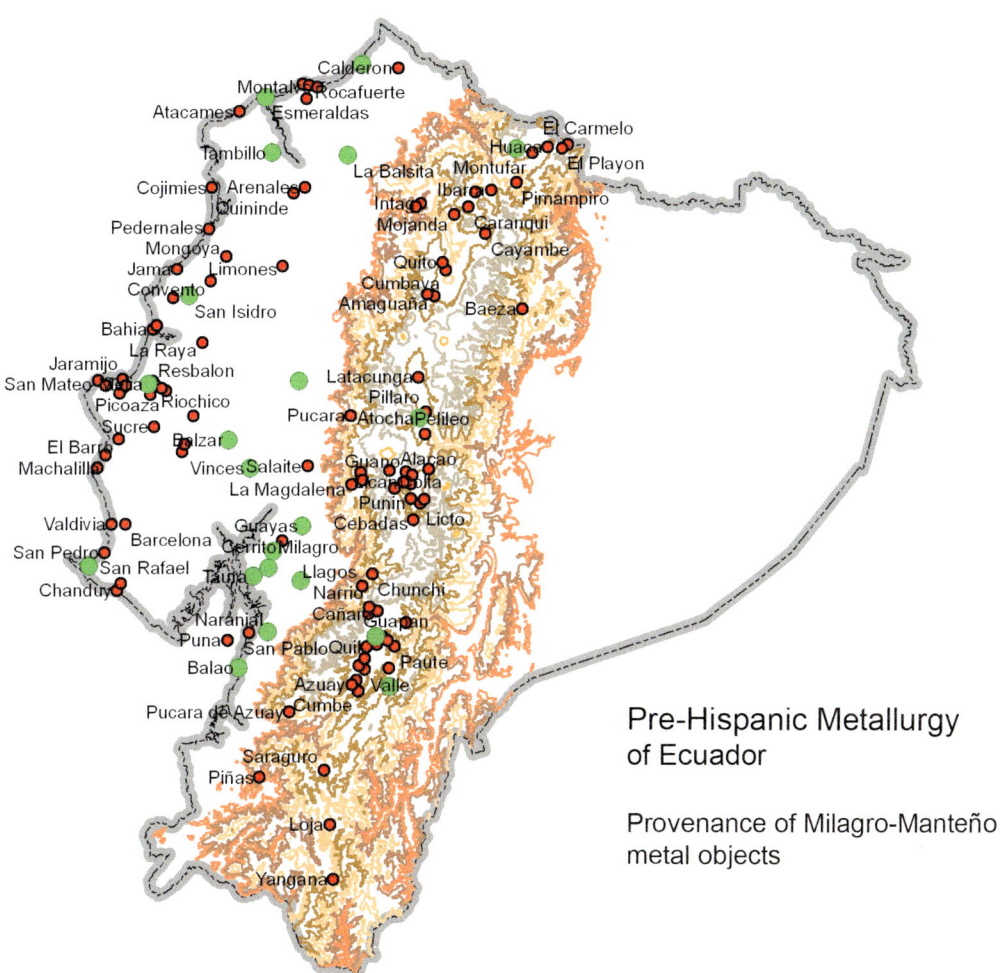

Pre-Hispanic Metallurgy of Ecuador

Provenance of Milagro-Manteño metal objects

2006) and the second from the Yumes site (Stemper 1993). In the following table the details of those dates are summarised:

Date	Canton	Laboratory Number	Bibliographic Reference
A.D. 120 ± 70	G-Guayas	WIS 1633	Stemper 1993
A.D. 1730 ± 60	G-Balzar	SI-171	Meggers 1966, R.D. 2006

The initial date from Yumes (Stemper 1993) seems to be too early; however its stratigraphic association is quite clear and it is not isolated; it is part of a sequence that covers from 2500 B.C. to A.D. 300 approximately (Stemper 1993) on several neighbouring sites. The presence of metallurgy in this site in the 2nd century A.D. is not surprising. It remains to be determined whether the objects belong to the Milagro – Quevedo proper style or if they represent some type of formative phase of it.

With relation to the final date for the 18th century (Meggers 1966 and R.D. 2006) it is important to recall that there are other similar dates for the region (Ubelaker 1981) that seem to point that until the end of the 17th or beginning of the 18th centuries the native funerary practices were still in vigour and that they included metal objects as offerings. The Spanish conquest took away gold and silver from the Indians but the manufacture of copper objects, as those found in La Compañía (Meggers 1966) and in Ayalan (Ubelaker, 1981), might have been permitted and could have persisted for a long time into the colonial period.

The available dates represent the probable extremes of the period of metal production that, according to the estimate of the Central Bank, would cover from A.D. 400 and A.D. 1533 (Pérez et al 1995). There are still many open questions related to possible phases and styles within the Milagro – Quevedo Group.

Technology

The wide varieties of metals and alloys present in Milagro-Quevedo metallurgy, as well as the massive use of copper, have been pointed out by several archaeologists and archaeo-metallurgists. Barrandon, Valdez y Estévez (2002) report the results of analyses of three tombac objects with compositions that range between 77.5 to 84.5% Au and 2.15 to 10.30% Cu; silver content (10.2 to 15.1%) is consistent with the natural alluvial percentage. Slight differences between surface and bulk composition suggest depletion gilding. However, gold objects in this Group are quite scarce and limited to small ornaments such as nose ornaments and ear pendants.

Copper and arsenical copper are the most important metals; they were used not only to make large objects and thousands of axe-monies but also for varied ornaments. The quantity of copper used, in a region that seems to lack natural deposits of this metal, implied structuring a network reaching supply sites that have not yet been identified.

Some copper objects were surface gilded; García et al (2000) confirmed by means of metallographic analyses the use of sheet gilding for a group of nose ornaments, some of them belonging to the Milagro – Quevedo Group. Scott (1985) examined a copper nose ornament, found in La Compañía, which had been plated with silver by fusion. Silver plating involved the following steps: 1) Shaping the nose ornament by hammering an ingot; 2) Cleaning the surface and possibly applying a flux; 3) Preparing an alloy made of approximately 40% silver and 60% copper; 4) Melting this alloy in a pottery crucible; 5) Immersing the copper nose ornament into the melted alloy; 6) Hammering and annealing to produce final shape and 7) Final cleaning. The two last steps would have corroded and depleted part of the copper thus enhancing the silver content (Scott 1985). The same author reports other objects from the region gilded or silvered by sheet plating.

Ubelaker (1981) found in Ayalan objects of gilded copper, together with evidences of welding with a flux containing lead. Holm (1970) mentions pottery beads overlaid with gold sheet. Zevallos (1965) reports the existence of depletion gilding and affirms that the technique was still in use some 40 years before by the peasants of Gualaceo that used a herb called chulco, boiled in salt water, to corrode surface copper and improve the appearance of the pieces.

McCall y Buchheit (1971) analysed 27 Milagro-Quevedo objects from La Compañía; metallographies confirm that they are copper – silver alloys (70 – 40% approximately) in a two phases structure. The objects, collar-button artefacts, were made from two parts, one of them cold worked from hammered and annealed wire, the other cast. In some cases the two parts were welded by the autogenous method (without flux), while in other cases welding was carried out with the help of an alloy with a lower melting point (close to the eutectic) and a flux. The technique is unusually complex for such small objects; it requires a very precise heat control and the ability to direct the flame to an exact point, something that was presumably achieved with blowers and whale oil. (McCall y Buchheit 1971).

Wire was hammered and annealed as the last step, to give it flexibility, then extensively used to make ornaments and instruments. Zevallos (1956) mentions objects made of shaped strips of wire over 2.70 ms. long. The width of the wire was carefully controlled, either keeping it uniform or making it thicker or thinner so that it fitted the final form of the ornament. Wire was used also to make needles that unlike those from Esmeraldas, belong to the punched eye type. The hole in these needles was made by drilling a widened and flattened end of the wire bending part of the flat part near the hole to strengthen the head of the needle (Holm 1963).

Sutliff (s.f., 1989, 1990) recovered in a domestic context in Peñon del Río a series of unfinished copper objects; her analyses allowed the reconstruction of the manufacturing process, made up of the following steps: 1) Casting ingots from previously refined copper; 2) Making pre-forms by casting the metal in the ingots; 3) Changing the size and shape of the pre-forms by hammering to obtain modified pre-forms; 4) Shaping and welding to obtain the final form and 5) Treating the surface to produce the final object (Sutliff s.f., 1989, 1990). These manufacturing strategies allowed the production of a wide variety of objects from a few artless pre-forms using only simple methods within domestic contexts (not specialised workshops) and for popular use (as opposed to elite use) (Sutliff s.f., 1989, 1990).

Casting was used simultaneously and in combination with hammering, as shown in the example of the collar-button artefacts. To cast some special objects, especially axes, moulds were used, some of them made from copper; the moulds are of the bivalve type (even though generally only one of the halves survives). We might suppose that these objects, as well as crucibles, must have been made in high melting point alloys, because otherwise they would have melted, together with their contents. Crucibles are circular and have three spillways (object 2-16-1, for example); inside them there are residues of what seems to be slag. Axes, weapons and tools were studied by Mayer (1992) who concludes that they were cast in univalve, bivalve or composite moulds (depending on their size or/and shape) and that afterwards most of them were cold worked to give them their final shape and harden the cutting edges and impact surfaces (except for giant axes that were not hammered). The decorative motifs present in some specimens were engraved (chiselled and pierced).

Axe-monies and related forms ("skins" and "emblems") studied by Hosler et al (1990) and Hosler (1998) deserve a separate mention in relation with their technology. These objects, that come from the regions of Milagro – Quevedo and Manteño – Huancavilca, are almost exclusively made from arsenical copper. There is nothing remarkable with respect to their manufacture; they were made by hammering from ingots or pre-forms, and they have a thick edge that gives them some strength. The technological challenge has to do with the volume of production: as opposed to other types of objects found in archaeological contexts axe-monies were deposited in large quantities. Up to 13,000 have been found in one site and it is usual to find them by hundreds inside pottery vessels or funerary urns (Hosler et al 1990). They are thin objects, of diverse sizes, with no apparent functional use, that required for their manufacture a centralised organisation to guarantee the supply of homogeneous raw materials and the production of uniform and patterned forms and sizes.

There are also examples of lost wax casting. This type of casting was used both in its basic version and with nucleus to make hollow objects such as the lime containers (object 68-21-14, for example). This process involves making a nucleus of charcoal and clay that was covered with wax; the nucleus was held in place with a support firmly attached both to it and the outer mould, thus assuring that the nucleus would not move during the melting down of the wax or the casting. In the case of lime containers this support was located in what was to be the mouth of the object, thus making it unnecessary to do any repairs when it was removed once the cast object was taken out of the mould.

Apart from collar-button artefacts welding is frequent in the manufacture of other gold, silver and copper objects. It was applied as small points to secure the internal endings of objects made with shaped wire or to join two wire spirals together. In small gold objects granulation welding was used to adhere small spheres forming rows and circles. Eutectic welding was employed to make globular necklace beads joining two hammered semi-spheres (Zevallos 1965).

The same techniques used for gold and copper were applied to silver, and identical objects were made. In a sense there was a continuum of silver – copper alloys; the objects having more copper appear as silvery copper (i.e. the collar-button artefacts analysed by McCall y Buchheit, 1971) while the objects with more silver appear as coppery silver (objects 62-1- 23 and 24, for example). In some cases there is an evident surface enrichment of silver that might have been intentional or the accidental result of alternate hammering and annealing (Sutliff 1989). Meggers (1966) reports finding bi-metallic objects that included silver components, pyrite mirrors with silver frames, crowns with real feathers and silver and gold feathers and textiles covered with silver plaques.

Stemper (1993) provides analyses of the objects found in the River Daule basin that confirm the use of copper – silver alloys; the author affirms that the results suggest that most are natural, not intentionally obtained alloys (Stemper 1993). This would mean that a silver mineral containing copper was used and that the copper survived even the refining process. The presence of mixed deposits of copper and silver in the poli-metallic province of the Andes would account for such finds (Oyarzún 2000).

As in La Tolita – Tumaco, in the region of Milagro – Quevedo lead spheres have been found occasionally; those found by Ubelaker (1981) in Ayalan are approximately 15 mms. in diameter and they have tombac inlays.

Decoration techniques include repousse, used mainly in laminar gold objects by hammering on both sides. Semi-precious stones like jadeite and turquoise were used as danglers or inlays. Spondylus was carved to make beads and inlays in copper objects. The finishing quality of Milagro - Quevedo objects is uneven; gold, silver and gilded copper pieces are generally well polished and there are no marks of hammering or casting burrs. Large copper objects, on the other hand, show a rough surface with marks of the casting moulds and hammering traces.

Zevallos (1965) reported a grave that he excavated in Guayas where he found the trade tools of a goldsmith including crucibles, chisels, copper pipes with holes to adapt blowers, conical "breads" of clay use to polish and tweezers. This exceptional find suggests that the trade of the metal Smith was clearly differentiated within this society.

Typology and classification

The repertoire of forms and functions of Milagro – Quevedo metallurgy suggests an important change in the trend of this industry in the coast. The technological changes described above have correspondent changes in the typological categories, as can be observed in the following table that summarises the characteristics of the simple in the collection of the Ministry of Culture:

Function	Form - Representation	Frequency	Percentage
Needle	Simple - total	14	4.31
Fishing hook	Simple – total	14	4.31
Shank	Simple – total	6	1.85
Rattle	Simple	12	
	Miniature	2	
	Total rattles	14	4.31
Chisel	Simple	17	
	Prismatic	5	
	Cylindrical	1	
	Shaped as a bird	1	
	Total chisels	24	7.38
Dangler	Simple - total	6	1.85
Bead necklace	Tubular - total	1	0.31
Crucible	Simple - total	1	0.31
Bead	Simple	1	
	Globular	2	
	Total beads	3	0.92
Disc	Simple - total	2	0.62
Figure	Anthropomorphic - total	1	0.31
Axe	Simple	4	
	With anthropomorphic figure	6	
	Shaped as a head	2	
	Shaped as a head, double	1	
	Shaped as a bird	1	
	With straight heel	5	
	Zoomorphic	1	
	With drilled heel	4	
	Total axes	24	7.38
Axe money	Simple - total	53	16.31

Function	Form - Representation	Frequency	Percentage
Sheet	Simple	2	
	Circular concave	1	
	Conical	2	
	Total sheets	5	1.54
Small mask	Miniature - total	2	0.62
Mould	Elongated - total	1	0.31
Nose ornament	Simple	80	
	With ornament	1	
	Zoomorphic shaped as an owl	1	
	Miniature	1	
	Shaped as an S	1	
	Spiral	14	
	Compound forma	1	
	Circular	1	
	Bevelled	1	
	Total nose ornaments	101	31.08
Ear pendant	Spiral - total	4	1.23
Breastplate	Simple - total	2	0.62
Tweezers	Simple – total	29	
	Zoomorphic shaped as an owl	1	
	Total tweezers	30	9.23
Container/bowl	Simple - total	1	0.31
Tumi (pin to hold robe)	Simple	10	
	With short heel	1	
	With wide heel	2	
	Total tumis	13	4.00
Tupo (traverse knife)	Simple - total	3	0.92
Total		325	100.00

Percentages in the left column indicate a quantitative distribution characterised by two predominant categories (nose ornaments with 31.08% and axe-monies with 16.31%) that represent almost half of the simple; there is a relatively uniform distribution of the other categories, none of which exceeds 10% of the sample. In the Milagro – Quevedo Group face decoration was limited almost exclusively to nose ornaments; this category is both the most abundant and the most varied, though the spiral is overwhelmingly dominant. In contrast ear pendants are scarce and quite simple. Equally scarce are necklaces, beads and breastplates. Tupos, though scarce (>1%), appear in this Group thus revealing an important change in the aesthetic patterns of body decoration.

However, personal adornments, are not as remarkable as tools, instruments and utensils (needles, fishing hooks, chisels, crucibles, axes, moulds, tweezers, bowls and tumis) that account for 37.54% of the sample. We would have to consider the group of axe-monies

(16.31%), a category whose function continues to be enigmatic (Hosler et al 1990), but that does not belong to the universe of personal adornments. What comes clear from this simple exercise is that this is a system of metallurgic production mainly oriented towards needs different from personal ornament; nevertheless, the reduced size of the sample restricts this conclusion to an indicative reference. This explains the massive use of copper, even in the absence of a local supply.

In the collection of the Ministry of Culture there are rings, lime containers (with their corresponding pins), projectile points and staffs (emblems) that are not recorded as such in the inventory. Apart from the above mentioned categories other authors record the existence of the following objects in the Milagro – Quevedo metallurgy: additional shapes of axes including giant axes weighing over 20 kgs., star pointed maces, hoes, drills, projectile points and awls (Mayer 1992); copper teeth for combs (Von Buchwald 1918); teeth inlays, chains, knives (different from tumis) and metallophones (bells) (Estrada 1957); pyrite mirrors with silver frames, crowns, silver plaques for textiles and spear thrower hooks (Meggers 1966) and pre-forms in several stages of manufacture (Sutliff 1989).

Figure 31 Milagro – Quevedo copper crucible: 3.2 x 12 x 12 cms.

FIGURE 32 MILAGRO – QUEVEDO COPPER MOULD: 1.2 X 7.7 X 17.1 CMS.

Great Regional Groups: Milagro – Quevedo

Figure 33 Milagro – Quevedo copper staff: 30 x 11.1 x 2.8 cms.

Figure 34 Milagro – Quevedo copper axe-monies: 10.5 x 10 x 0.5; 9.2 x 8.3 x 05 and 10.9 x 10 x 0.5 cms.

Figure 35 Milagro – Quevedo gold nose ornament: 4 x 5.6 x 0.4 cms.

FIGURE 36 MILAGRO – QUEVEDO GOLD SPIRAL NOSE ORNAMENT: 2.6 X 1.4 X 0.1 CMS.

Chapter 9

Great Regional Groups: Manteño - Huancavilca

The Manteño- Huancavilca Group is one of the most important in Pre-Hispanic Ecuador. This is due, partly, to the legendary discovery of graves with impressive quantities of metal objects (Estrada 1957, Farabee 1921) and, also, to the role that it played in the diffusion of metallurgy to western Mexico (Hosler 1998). IN the Ministry of Culture this Group is represented by a modest sample.

Geographic Distribution

In the following table we summarised the available provenance information for the Manteño – Huancavilca Group in the collection of the Ministry of Culture (Quito):

Province	Canton	Number of objects	Percentage of total
Cañar	Cañar undefined	1	0.3
Unknown	Unknown	4	1.2
Esmeraldas	Quininde	1	0.3
Guayas	Chanduy	71	21.8
	El Cerrito	1	0.3
	San Pedro	19	5.8
	San Rafael	42	12.9
Manabi	Bahía	1	0.3
	Cerro de Hojas	1	0.3
	El Barro	30	9.2
	Manabí undefined	66	20.2
	La Raya	2	0.6
	Las Chacras	3	0.9
	Mejia	61	18.7
	Miguelillo	10	3.1
	Portoviejo	5	1.5
	Rió Mariano	1	0.3
	San Isidro	1	0.3
	San Mateo	2	0.6
	Salaite	4	1.2
	Total	326	100

The map shows the relation between the provenances for Manteño – Huancavilca objects (green dots) with relation to all the other provenances (red dots):

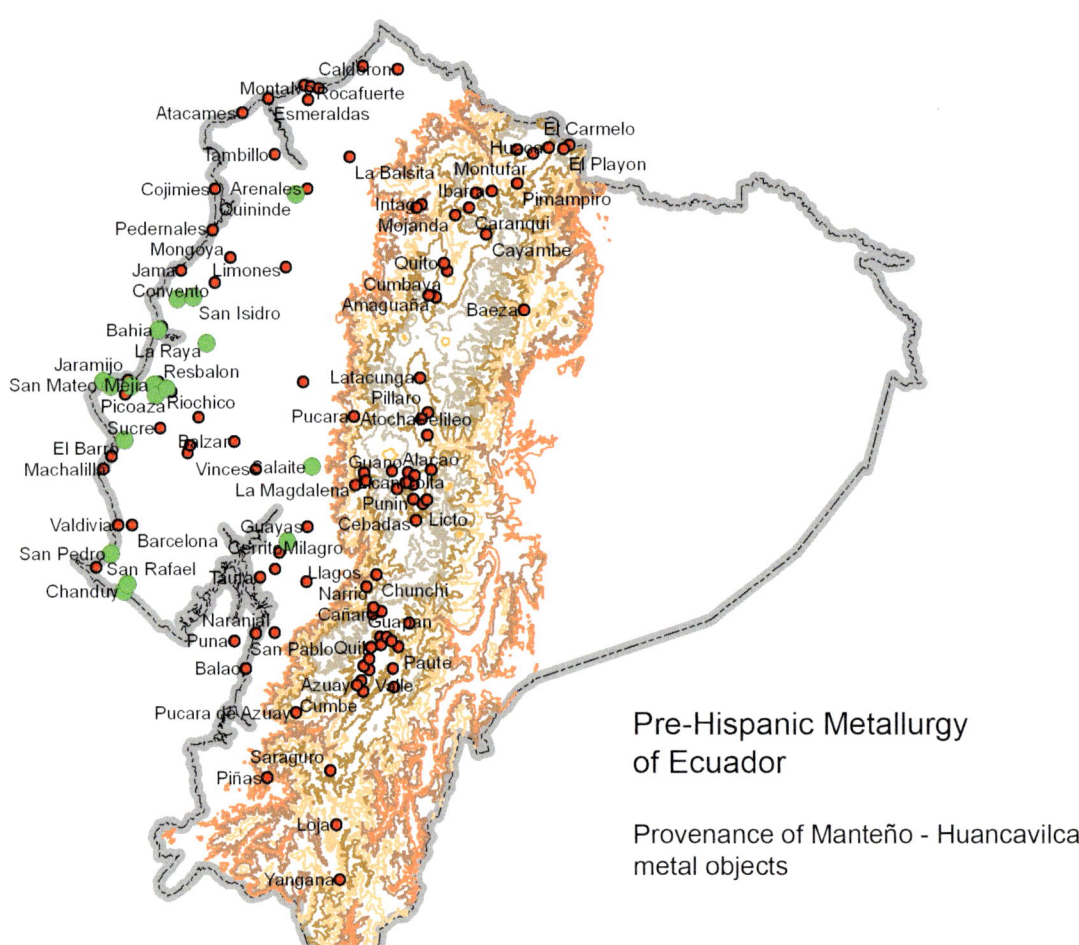

Pre-Hispanic Metallurgy of Ecuador

Provenance of Manteño - Huancavilca metal objects

Just over 21% of the objects lack provenance information or have a very deficient information. The useful simple to determine distribution patterns is therefore reduced to less than 79% that is 256 objects. Only one object seems to have been found outside the coastal area, in the province of Cañar, and it may be the product of exchange or a wrong provenance declaration.

In the coastal area there is only one object found north of Manabí (Quininde, Province of Esmeraldas). The rest of the finds come from the provinces of Manabí, Guayas and, to a lesser extent, Los Ríos. This distribution is coherent with the territory traditionally accepted the Manteño – Huancavilca territory. The concentration foci are in the areas of the Santa Elena peninsula (Chanduy, San Rafael, San Pedro) and around Portoviejo (El Barro, Mejia, Miguelillo, Las Chacras and Portoviejo). It would be interesting to compare the spatial correlation between metallurgy and vestiges of villages and public buildings in these two areas.

Chronology

Most of the dates directly associated to Manteño – Huancavilca metallurgy come from the excavations of Ubelaker (1981) in the cemetery of Ayalan. In the table the relevant data for the dates is summarised:

Date	Canton	Laboratory Number	Bibliographic Reference
A.D. 730 ± 115	G-Guayas	SI-3529	Ubelaker 1981
A.D. 800 ± 7	G-Guayas	SI-3532	Ubelaker 1981
A.D. 985 ± 70	G-Guayas	SI-3534	Ubelaker 1981
A.D. 1110 ± 60	G-Guayas	SI-3309	Ubelaker 1981
A.D. 1155 ± 90	G-Guayas	SI-3308	Ubelaker 1981
A.D. 1610 ± 40	M-El Barro	Beta 210356	Museo del Oro 2005
A.D. 1730 ± 60	G-Guayas	SI-3530	Ubelaker 1981

The sequence represented by these seven dates covers a time span of a thousand years from A.D. 700 and A.D.1700 approximately. Even though there is still a long period without dates (between the 13th and 17th centuries) the sequence seems to confirm the most accepted chronological estimate that goes from A.D. 600 to A.D. 1533 (Pérez et al 1995). It is likely that in such a long time there were more than one period or phase, but there is no information about it.

There are two issues that we need to reconsider since they are related to the initial and final dates of the sequence. Even though the proper Manteño – Huancavilca style is present around the 7th or 8th centuries we must remember that there are evidences of metallurgical activity exactly in this zone as early as 1500 B.C. (Hosler 1998) and A.D. 890 (Zevallos 1965). There is, therefore an obscure period lasting around 1,500 years (900 B.C. to A.D. 600) about which we know virtually nothing that corresponded to the time when the metallurgy of the Ecuadorian coast developed and formed differentiated styles. With relation to the final date we have the opinion of Ubelaker (1981) himself who believes that it is excessively late. We believe that traditional burials and copper metallurgy might have persisted in the Ecuadorian coast until the beginning of the 18th century.

Technology

Manteño – Huancavilca metallurgy is quite complex, both with relation to the metals and alloys used and the techniques used. It shares, however, with the Milagro – Quevedo Group several characteristics, techniques and processes: for this reason we shall omit the description of common themes that have already been discussed in the previous section.

Barrandon, Valdez y Estévez (2002) report three analyses for objects from this region; two of them are tombacs with very low contents of copper and silver (80.4%Au, 9%Ag, 8.9%Cu and 95.7%Au, 6.4%Ag, 4%Cu) while the third is a low silvery tombac (43.8%Au, 22.2%Ag, 33.6%Cu). This indicates, grosso modo, the use of tombacs with variable composition. Gold and tombac objects are, however, relatively scarce in the region in comparison to the copper, and even the silver finds. Silver, gold and copper, or gilded copper, appear combined in bi-metallic objects like silver masks with attached copper crowns.

Hammering is the most extensively used technique; the method was used to work gold, copper and silver. To obtain large and tri-dimensional objects they used several assemblage techniques; among them it is particularly interesting the assemblage of several sheets in the same plane by means of hooks to make complex breastplates and ear pendants (objects 71-25-26, 71-47-1 and 71-47-4, for example). Flanges were used to assemble small gold objects. Welding is present, especially in small gold objects; it is not too frequent and we are not sure whether it is eutectic welding or if they used welders and fluxes. There is also granulation welding in gold objects.

Hammering and shaping of long wires to make nose ornaments and ear pendants with spiral motifs is also frequent. The necessary complement of hammering that is cutting of sheets was made by means of chisels and was mastered to a high degree; many Manteño – Huancavilca objects have intricate shapes with indentations and prolongations that were symmetrically and delicately made, with no noticeable imperfections. Large laminar objects were usually decorated by repousse, working on both sides; there are some metallic instruments with blunt endings that were probably used for repousse (objects 69-36-16 to 20, for example). Some solid copper pieces were decorated by engraving.

Gold and copper needles with the hole made by the punched eye technique described for Milagro-Quevedo needles (Holm 1963) are quite frequent. Other common object, copper and gilded copper fishing hooks, thick and hardened by cold working, were provided with shackles by widening the shank and folding it on itself.

There are evidences of depletion gilding for tombac objects and other types of gilding (fusion and sheet) for copper (García et al 2000); gilded copper is, in fact, very frequent. Probably they had also silver plating for copper and silver – copper alloys. For decorative purposes they used inlays and danglers of semi-precious stones like turquoise and lapis lazuli, the later one brought from Peru or Chile.

In spite of the fact that hammering was the main technique, casting seems to have been quite important. I is not only that there are many small cast copper figures but also that many of the large hammered objects had to be cast to make a pre-form before finishing by hammering. This is the case with axes and knives; Marcos (1995) reports the existence of open moulds for this type of objects, together with pre-forms for axe-monies in the site of El Cangrejito. Holm (1970) recorded the presence of large kilns for casting in the coast of Manabi.

Axe-monies are as abundant in the Manteño – Huancavilca Group as they are in the Milagro – Quevedo, or even more; forms, manufacturing techniques and alloys are virtually indistinguishable. The metallurgic industry in this region reached such a degree of development that it triggered a diffusion process that reached western Mexico (Hosler 1998). Manteño metal smiths traded over long distances and maintained a work equipment that included weighing scales of the "roman" type to weigh the metals (Bray 1971).

Typology and classification

The geographic proximity, that in certain regions is really an overlap, between the Milagro – Quevedo and Manteño – Huancavilca Groups has determined that in many instances they come to be treated together and assumed to share the same forms and functions of their objects. This approach must be carefully weighed in order to study in depth the particularities of both Groups. In the following table we summarised the available information about form and function of Manteño objects in the collection of the Ministry of Culture (Quito):

Function	Form - Representation	Frequency	Percentage
Sub-labial ornament	Simple - total	1	0.34
Needle	Simple - total	12	4.05
Wire	Simple	2	
	Annular	2	
	Spiral	1	
	Total wires	5	1.69

Function	Form - Representation	Frequency	Percentage
Fishing hook	Simple - total	1	0.34
Shanks	Simple	7	
	Open	2	
	With stone	1	
	Total shanks	10	3.38
Staff	Simple - total	3	1.01
Burin	Simple - total	1	0.34
Lime container	With handle - total	1	0.34
Rattle	Simple	17	
	Semi-lunar	1	
	Conical	1	
	Total rattles	19	6.42
Chisel	Simple	5	
	Bevelled	1	
	Total chisels	6	2.03
Headband	Simple - total	2	0.68
Dangler	Tinculpa type- total	1	0.34
Bead necklace	Simple	3	
	Oval	1	
	Double	1	
	Miniature	1	
	Total bead necklaces	6	2.03
Bead	Oval - total	3	1.01
Disc	Simple - total	2	0.68
Spatula	Simple - total	6	2.03
Figure	Head shaped with ornament	1	0.34
Axe	Simple	4	
	With anthropomorphic figure	3	
	With straight heel	23	
	With short heel	1	
	Zoomorphic	1	
	Total axes	32	10.81
Axe-money	Simple	73	
	With straight heel	1	
	With short heel	1	
	Total axe - monies	75	25.34
Sheet	Simple	4	
	With anthropomorphic figure	1	
	Annular	2	
	Zoomorphic	1	
	Total sheets	8	2.70
Mask	Anthropomorphic - total	2	0.68

Function	Form - Representation	Frequency	Percentage
Nose ornaments	Simple	3	
	Head shaped zoomorphic	1	
	Annular	4	
	Circular	1	
	Double spiral	1	
	With stone	1	
	Cylindrical	1	
	Shaped as a U	1	
	With spiral endings	1	
	Spiral	1	
	Total nose ornaments	15	5.07
Ear pendants	Annular with rim	10	
	With ending	4	
	Spiral	5	
	Spool shaped	1	
	Wire spiral	2	
	Total ear pendants	22	7.43
Breastplate	Simple	2	
	Tinculpa type	5	
	Square	3	
	Zoomorphic Tinculpa type	2	
	Zoomorphic	3	
	Head shaped zoomorphic	1	
	Head shaped anthropomorphic	1	
	Circular	4	
	Anthropomorphic Tinculpa type	3	
	Total breastplates	24	8.11
Tweezers	Simple	10	
	Tubular	3	
	Circular	4	
	Total tweezers	17	5.74
Container/bowl	Simple - total	3	1.01
Drill	Simple - total	10	3.38
Tumi	Simple	4	
	With anthropomorphic figure	1	
	Double	2	
	Total tumis	7	2.36
Tupo	Simple - total	1	0.34
Total		296	100.00

Adornments to be worn on the head and face have in this Group a much lower relative importance than in any other of the coastal Groups; headbands, sub-labial adornments, masks, nose ornaments and ear pendants add up only to 14.2% of the sample. In contrast the elements to be worn on the chest (bead necklaces, beads, danglers, tupus and breastplates), that are scarce in the other Groups have in Manteño-Huancavilca more importance (11.83%) thus balancing the proportion of ornaments for heads and chest. Breastplates, the most important type of ornament for the chest, are varied and carefully manufactured.

The group of utensils, instruments and tools is also very important; needles, fishing hooks, burins, lime containers, chisels, spatulas, axes, tweezers, bowls, drills and tumis add up to almost one third of the sample (32.43%). If we were to add the axe-monies (25.34%) we would have a total of 57.77% of objects that are not body ornaments. Even though this is consistent with the pattern found for the same group of objects in Milagro – Quevedo (53.85%), the relative frequencies of the diverse categories varies considerably from one Group to the other. The same can be said of personal ornaments.

Even though they are not recorded as such in the inventory, in the collection of the Ministry of Culture there are: a box completely sealed by welding, double and triple tinculpas, nipple covers, one piece necklaces, crowns and silver plaques; some of the later ones might have been used to be sewn on textiles while the larger ones could have been adhered to the walls of buildings. Apart from the objects already mentioned there are reports about: axes of various shapes, knives different from tumis and projectile points (Mayer 1992); "skins" (Hosler et al 1990) and face nails, rings and helmets (Pérez et al 1995).

Great Regional Groups: Manteño - Huancavilca

Figure 38 Manteño – Huancavilca silver and copper mask with crown: 30.5 x 18.3 x 15.3 cms.

FIGURE 39 MANTEÑO – HUANCAVILCA SILVER AND COPPER MASK WITH CROWN:
9 x 7.9 x 6 CMS.

Figure 40 Manteño – Huancavilca silver breastplate: 23.3 x 23 x 0.9 cms.

FIGURE 41 MANTEÑO – HUANCAVILCA SILVER PLAQUE: 13 X 31 X 0.1 CMS.

Great Regional Groups: Manteño - Huancavilca

Figure 42 Manteño – Huancavilca copper axe: 14 x 16.3 x 0.6 cms.

FIGURE 43 MANTEÑO – HUANCAVILCA COPPER BREASTPLATE, TINCULPA STYLE:
21.4 x 5.5 CMS.

Chapter 10

Great Regional Groups: Puruha

The Puruha Group is one of the less known metallurgical traditions, in spite of the discovery of rich graves with abundant metal objects since colonial times. In the collection of the Ministry of Culture the Puruha Group is represented by a relatively small sample.

Geographic Distribution

The geographic distribution of the objects of the Puruha Group is summarised in the following table:

Province	Canton	Number of objects	Percentage of total
Unknown	Unknown	29	11.2
Azuay	Azuay	1	0.4
	Cuenca	3	1.2
Bolívar	Conventillo	8	3.1
Chimborazo	Alacao	137	53.1
	Chimborazo	16	6.2
	Chunchi	2	0.8
	Guano	4	1.6
	Lican	3	1.2
	Llagos	16	6.2
	Molobog	25	9.7
	Chimborazo undefined	4	1.6
Cañar	Pindilig	1	0.4
Cotopaxi	Latacunga	5	1.9
Tungurahua	Pillaro	4	1.6
	Total	258	100

In the following map (page 112) the distribution of Puruha objects (Green dots) is displayed in relation to the distribution of objects from all the other Groups (red dots); collection of the Ministry of Culture (Quito).

The distribution of the metal objects of the Group is consistent with the territory of the archaeological culture and the Puruha historic ethnic group in the 16th century. The southern limit of this distribution is in Azuay (Azuay and Cuenca) while the northern limit reaches southern Cotopaxi (Latacunga). There are extensions to the west, province of Bolívar (Conventillo,) that are within the expected pattern. The

Pre-Hispanic Metallurgy of Ecuador

Provenance of Puruha metal objects

concentration focus is in the province of Chimborazo (Alacao, Molobog, Llagos, Chimborazo, Guano), the sites where presumably were the centres of political power of the Puruha chiefdoms before the advent of the Incas. In spite of the coherence of the pattern it is worth warning that the sample is small and therefore statistically weak and that the acquisition of large sets (from Alacao and Molobog) might induce some distortion.

Chronology

Up to date we have four dates associated with Puruha metallurgy. All of them come from collection objects without a detailed archaeological context:

Date	Canton	Laboratory Number	Bibliographic Reference
A.D. 210 ± 40	CH-Molobog	Beta 237172	Museo del Oro 2007
A.D. 240 ± 40	CH-Alacao	Beta 237170	Museo del Oro 2007
A.D. 340 ± 40	CH-Alacao	Beta 210354	Museo del Oro 2005
A.D. 860 ± 40	CH-Molobog	Beta 210355	Museo del Oro 2005

The period of approximately 650 years comprised by these dates differs considerably from the chronological estimate of the Banco Central that goes from A.D. 1200 to A.D. 1533 (Pérez et al 1995). We must recognise, however, that the terminal date for the autonomous Puruha metallurgy does coincide in the central Ecuadorian Sierra with the Inca expansion in the 15th century; from then onwards the metallurgical industry is integrated or absorbed by the Incas. On the other hand the initial date for metallurgy in this region is definitely earlier than the 13th century A.D.

As already discussed, metal objects were being manufactured since 1500 B.C. in workshops in the south Sierra and its diffusion to the central Sierra must have been quite rapid. The date for the 3rd century (A.D. 210) is not at all excessively early; in fact, it is associated with a perfectly established metallurgical tradition, earlier phases probably exist. A more coherent estimate would place the development of Puruha metallurgy between A.D. 200 and A.D. 1500. During this period there were possibly two or more different phases; those have not been identified in the material available in the collections. The absence of archaeological research in the area prevents the construction of more precise and detailed chronologies.

Technology

Apart from a few isolated analyses that we shall mention later, there have been no systematic studies of Puruha metallurgical technology. On the basis of the examination of the collection it is evident that in Puruha gold, silver and copper were extensively used, either alone or alloyed. Barrandon, Valdez y Estévez (2002) carried out analyses on four Puruha objects by NAA (matrix composition) that turned out to be a silvery gold (84.6%Au, 15.3%Ag), two tombacs with low copper content (5.1 y 12.9% Cu respectively) and a tombac with high copper content (44.6%Au, 48.6%Cu). Once again the conclusion is that silvery gold was used and alloyed with variable quantities of copper; the result is a wide range of tombacs. The examination of colour surface of several objects of the collection confirms that possibility since there are many shades of yellow that correspond to different contents of copper. XRF (surface composition) analyses of the same objects shows higher proportions of gold and lower of copper, something consistent with surface enrichment; it is not yet possible to establish if the enrichment was intentional or the result of manufacture and /or burial conditions.

Copper and silver are present in noticeable quantities in Puruha metallurgy. Copper is present with or without gilding; in this last case the probable techniques are fusion or sheet gilding, but we do not know yeti f only one or both techniques were used. Rivet y Arsandaux (1946) mention two copper objects from Jordán and Chumbe with silver plating; the technique would be the one described by Scott (1985).

In order to understand the composition and manufacturing technique of some Puruha objects, XRF analyses and metallographies were done on four samples from Alacao and Molobog in the province of Chimborazo, acquired between 2003 and 2005 (DTI 2006). A dangler and its corresponding hook turned out to be tombacs of different composition produced on the basis of the same type of silvery gold; metallographies revealed cold working structures with surface gilding. A fragment of a staff cover is a silver – copper alloy; the structure shows segregated bands oriented in the direction of mechanical working. A fragment of dangler is also a tombac made by alloying silvery gold with copper and gilding the surface; there are evidences of casting followed by cold working. In a much mineralised sample of copper here were traces of a gold layer (DTI 2006). The evidences point towards a predominance of mechanical working and the extended use of surface gilding.

In quantitative terms hammering and its auxiliary techniques are, therefore, more frequent than casting. Nevertheless, there are several cast figures and, sometimes, cast and hammered components are combined in one piece (objects 5-2-5 y 6, for example). Casting probably started with gold from mines (Barrandon, Valdez y Estévez 2002) and refined copper minerals; there are a few gold ingots used as offerings in the graves. Casting techniques included the lost wax method for figures, spear thrower hooks and small objects and mould casting for larger objects like axes.

Hammering and annealing were very well controlled; in this Group gold and silver very thick hammered objects are frequent, even though there are also very thin sheets. Cutting was done with chisels; on thick objects it is usual to find rectangular cuts that constitute a distinctive mark of this metallurgical tradition. Hammered wire was employed to make complex pieces by shaping and assemblage (object 2-5-16, for example).

Assemblages are quite complex and they involve mechanical methods like hooks and slots, rivets that join together sheets and cast components (objects 5-2-5 y 6) and shackles. Welding was employed to fix shackles onto sheets and adhere danglers (usually in very large tupos), to join together two halves of rattles, to join shanks to endings in the tupos, to adhere sheets and wires and to fix mechanical assemblages made by shaping wire (object 67-5-67, for example). We do not yet know whether it is autogenous welding or if fluxes were used.

There is a type of assemblage that has not been studied by metallographic methods yet; it is extremely interesting. It is a technique used to make nose ornaments and ear pendants (objects 5-2-24, 25 y 30, for example); those are thick laminar objects made by hammering, most of their surface is gold but one or several segments are silver inlays that appear on both surfaces.

Preliminarily we can suggest that they were made as follows: 1) The starting point of the process consists in casting two pre-forms, one of a gold alloy and the other a silver alloy with equal or very close melting points; 2) The pre-forms were hammered until they were just a little thicker than the final thickness desired; 3) The inner surfaces of the gold and silver pre-forms were worked by cutting and abrasion until they fitted together perfectly; 4) By means of the welding technique known as exudation, that consists in heating the surfaces to be welded until the metal acquires a viscous state and then pressing the surfaces together, the silver inlays were adhered to the gold pre-forms; 5) The bi-metallic pieces were then cold worked and annealed until they reached the desired thickness. This process would have improved the adherence by propitiating solid inter-diffusion in the contact zone. Also the thickness of the gold and silver components would have been levelled; 6) The final stages would have been to drill the slots for the hooks, decorate by engraving and polish the edges by abrasion. There was yet another technique used to make objects with two different colours on the surface. It constitutes a variant of zoned scraping and is described in the following chapter dedicated to the Cañari Group because it is more frequent in there.

Decorative techniques include repousse, use by itself or combined with engraving. The later technique is much more frequent than repousse and used on objects made of all metals and alloys; chisels with their cutting edges hardened by cold working were used for engraving and the resulting designs are quite complex and intricate.

Metal plating of wooden staffs and spear throwers is quite remarkable. Usually the objects are large (1.50 ms. or more) and required using several gold or silver sheets assembled together over the wood core. Sheets were embossed reproducing the carving on the wood underneath. To fix the plating small gold and silver nails were used.

Typology and classification

Puruha metallurgy is one of the least known in Pre-Hispanic Ecuador; it is not only that there are no archaeological studies that might reveal the burial contexts but also that, unlike Cañar and Azuay, the grave lootings were not even documented. This is why the study of form and function starts almost from zero. In the following table we summarise the forms, functions and the frequencies of the Puruha metal objects in the collection of the Ministry of Culture:

Function	Form - Representation	Frequency	Percentage
Needle	Simple - total	1	0.4
Wire	Simple - total	4	1.6
Ring	Simple	1	
	Anthropozoomorphic	1	
	Wire spiral	1	
	Total rings	3	1.2
Shank	Simple	9	
	With extension	1	
	Of a larger object	2	
	Total shanks	12	4.7
Staff	Simple - total	2	0.8
Button	Globular rattle - total	1	0.4
Bracelet	Simple - total	3	1.2
Rattle	Simple	4	
	With spiral endings	4	
	Fusiform	1	
	Total rattles	9	3.5
Helmet	Simple – total	3	1.2
Headband	Simple - total	1	0.4
Pendant	Shaped as a head Anthropomorphic	2	
	Triangular with dangler	1	
	Circular Anthropomorphic	1	
	With dangler	4	
	Tinculpa type	2	
	Circular	5	
	Hollow	2	
	Shaped as a head with dangler	1	
	Total danglers	18	7.0

Function	Form - Representation	Frequency	Percentage
Bead Necklace	Simple	1	
	Miniature	1	
	Total bead necklaces	2	0.8
Bead	Simple	3	
	Circular	9	
	Tubular Anthropomorphic	1	
	Triangular with dangler	3	
	Of a larger object	2	
	Triangular	3	
	With dangler	2	
	Oval	2	
	Total beads	25	9.7
Diadem	Simple	1	
	Shaped as a head anthropomorphic double	1	
	Total diadems	2	0.8
Figure	Shaped as a bird - total	1	0.4
Spear thrower hook	Simple	1	
	Anthropomorphic	1	
	With zoomorphic figure	1	
	Total spear thrower hooks	3	1.2
Axe	With holes	1	
	With drilled heel	34	
	With openwork heel	1	
	With straight heel	1	
	Total axes	37	14.3
Sheet	Simple - total	1	0.4
Nose ornament	Simple	10	
	Tubular	1	
	With spiral endings	4	
	Oval	1	
	Shaped as a head anthropomorphic	1	
	Cylindrical	1	
	Circular	1	
	Elliptical	2	
	Elliptical with horizontal extensions	3	
	Spiral	1	
	With dangler	2	
	Total nose ornaments	27	10.5

Function	Form - Representation	Frequency	Percentage
Object	Oval	2	
	Shaped as a band	1	
	Shaped as a double band	1	
	Circular	2	
	Simple	2	
	Total objects	8	3.1
Ear pendant	Simple	6	
	With zoomorphic figure	5	
	Cylindrical with bead	2	
	With bead	12	
	Double	1	
	Circular hollow	12	
	Circular	1	
	With dangler	1	
	Spiral	4	
	Cylindrical	20	
	Shaped as a band	2	
	Total ear pendants	66	25.6
Breastplate	With wide endings	2	
	Rectangular	2	
	Square	1	
	Circular	2	
	Total breastplates	7	2.7
Nipple cover	With an accessory on the tip - total	2	0.8
Various objects	Total	1	0.4
Spear thrower	Simple - total	4	1.6
Head breaker	Star shaped - total	1	0.4
Tumi	Simple - total	1	0.4
Tupo	Simple	8	
	With spiral ornament	2	
	With shank	1	
	With anthropomorphic figure	1	
	Anthropomorphic double	1	
	Total tupos	13	5.0
Total		258	100.00

This Sierra Group repeats, though in a moderate scale, the pattern observed in the coast Groups that emphasises on the decoration of the face and head over the rest of the body. The set formed by helmets, headbands, diadems, nose ornaments and ear pendants represents 38.5% of the sample. Nevertheless, there is a certain balance between this area and the chest since the set of danglers, bead necklaces, beads, nipple covers and tupos adds up to 26.0%. The remaining body ornaments for arms and hands (rings, shanks and bracelets) adds up to only 7.1%. Utensils, tools and instruments, including staffs, spear thrower hooks, spear throwers, axes and tumis make up 18.3% of the collection.

The most varied and elaborate types of objects are ear pendants, pendants and tupos. Ear pendants are, sometimes, very large, profusely decorated and with danglers that make them real rattles; pendants are similar in that respect and, even though they do not tend to be large, they are carefully and finely made. Breastplates, on the contrary, tend to be simple and plain (there are a few embossed) and quite large. The largest objects, however, are tupos; some of them are more a metre long, thus indicating that they were meant only for funerary purposes, because such a large adornment would be too heavy and awkward to use (Ontaneda y Fresco 2002).

Staffs and spear throwers are, or were, mounted on a wooden nucleus that occasionally has survived. The gold, silver and copper embossed sheets that make them were assembled to the wood with nails; this assemblage was strong enough to hold solid gold or copper hooks.

Even though they are not recorded as such in the inventory, in the collection of the Ministry of Culture there are copper and silver nails used for unknown purposes, whistling projectile points, silver bowls and copper crowns with extensions that resemble feathers.

Figure 45 Puruha gold spear throwers: 45 x 1.4; 46 x 1.4 and 45 x 1.8 cms.

Great Regional Groups: Puruha

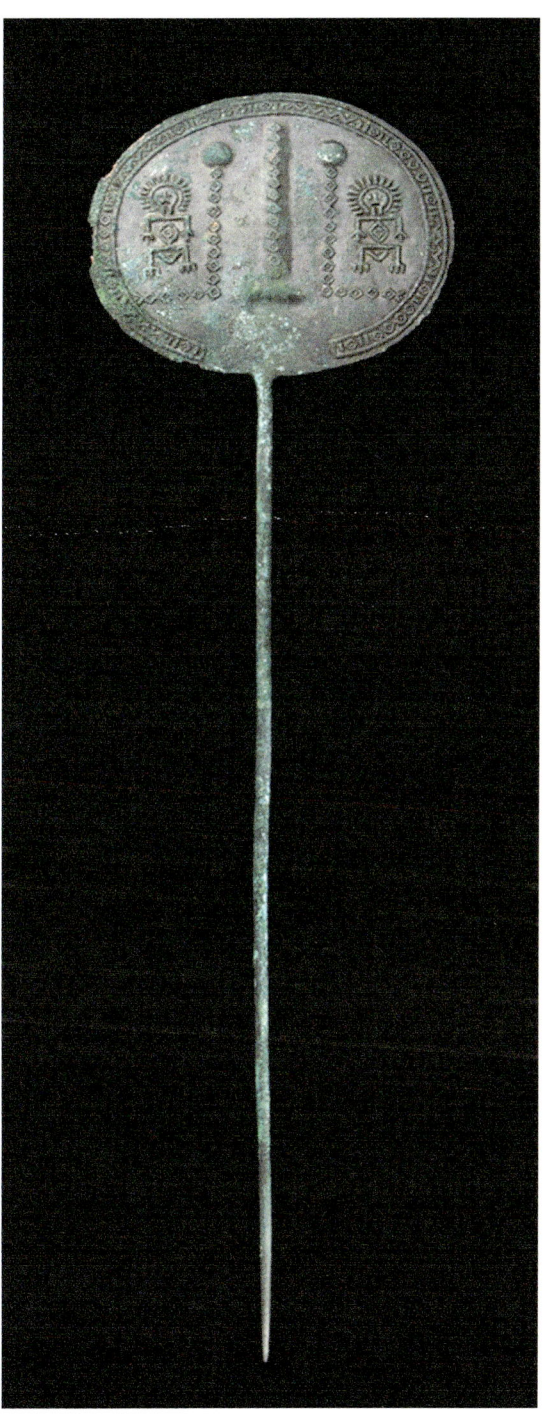

Figure 46 Puruha giant copper *tupo*: 45.8 x 13.5 x 1 cms.

FIGURE 47 PURUHA COPPER CROWN: 21 X 16.3 CMS.

Great Regional Groups: Puruha

Figures 48 and 49 – Puruha gold ear pendants with zoomorphic figures: 7.2 x 6.5 x 2.9 and 6.7 x 6.7 x w2.9 cms.

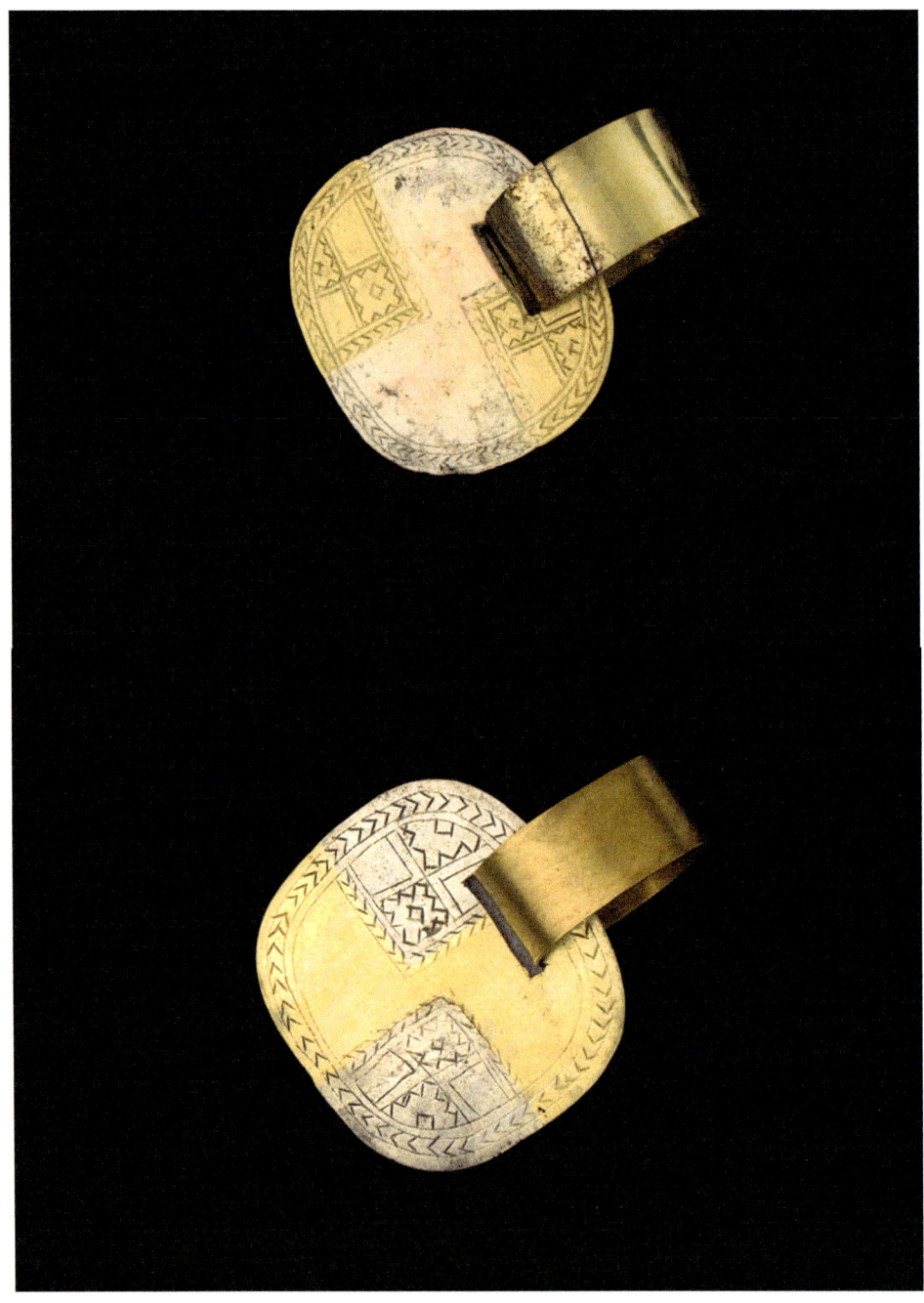

FIGURES 50 AND 51 – PURUHA GOLD AND SILVER EAR PENDANTS:
3.7 x 3.5 x 1.1 AND 4 x 3.6 x 1 CMS.

Figure 52 Puruha gold and silver nose ornament: 5.1 x 5.8 x 0.2 cms.

Figure 53 Puruha gold anthropomorphic pendant: 4.2 x 4 x 1 cms.

Chapter 11

Great Regional Groups: Cañari

The Cañari Group is much less known and scarcer than what would be presumed from the extravagant chronicles of the Spanish conquistadors. It seems that the finds of the rich graves of Guapan, Cuenca, etc. (Uhle 1922, Heuzey 1870) are only in the past and certainly not corresponded by the contents of present day collections.

Geographic Distribution

In the following table we summarise the provenance information for the small sample of Cañari metallurgy in the collection of the Ministry of Culture (Quito):

Province	Canton	Number of objects	Percentage of total
Unknown	Unknown	1	0.9
Azuay	Chordeleg	27	24.1
	Huaynacapac	1	0.9
	Pucara de Azuay	48	42.9
	Quil	1	0.9
	Yanuncay	1	0.9
Chimborazo	Cebadas	4	3.6
Cañar	Cañar	3	2.7
	Guapan	1	0.9
	Narrio	6	5.4
Imbabura	Caranqui	19	17.0
	Total	112	100

The finds of Caranqui in the Imbabura province, representing a significant percentage of the sample (17%), are extremely surprising and might correspond to a wrong provenance declaration rather than to the result of pre-Hispanic exchange. The rest of the provenances shape a coherent pattern covering the provinces of Azuay, Cañar and southern Chimborazo. As a matter of fact it is surprising not to have a larger distribution area, though this is due probably to the reduced size of the sample. In the map the provenances of Cañari objects (green dots) are displayed in relation to the provenances of objects from all the other objects of the collection of the Ministry of Culture (red dots):

Metallurgy in Ancient Ecuador

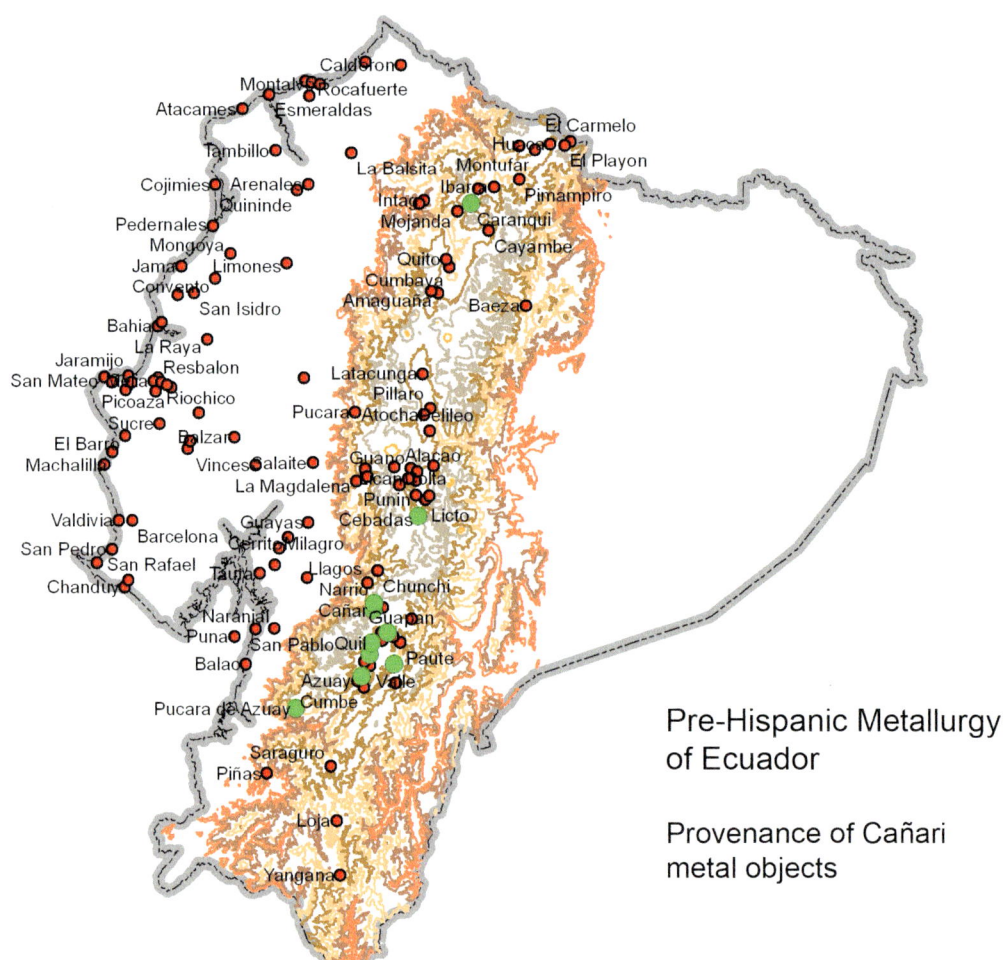

Pre-Hispanic Metallurgy of Ecuador

Provenance of Cañari metal objects

Chronology

All absolute dates for Cañari metal objects are the result of the excavations of the Spanish Mission in Ecuador and specifically of the digs in Ingapirca (Fresco 1984). In the following table the basic information on radiocarbon dates associated to metallic objects is summarised:

Date	Canton	Laboratory Number	Bibliographic Reference
A.D. 990 ± 70	CÑ-Ingapirca	CSIC-319	Fresco 1984
A.D. 1260 ± 80	CÑ-Ingapirca	CSIC-322	Fresco 1984
A.D. 1370 ± 70	CÑ-Ingapirca	CSIC-323	Fresco 1984

The chronological estimate of the Banco Central goes from A.D. 500 to A.D. 1533 (Pérez et al 1995); in this sense the Ingapirca sequence would cover just an intermediate segment of the period. However, it is important to check this assumption against the radiocarbon dates and the ethno-historic information for the area. As previously discussed, in Putushio just south of the Cañari area, there is a sequence that covers the span from 1470 B.C. to 865 B.C. (Rehren y Themme 1994). It is quite feasible that metallurgic activity would have continued after this last date and that it expanded rapidly on the neighbouring regions like Azuay and Cañar. The tracks of the metallurgy produced between 800 B.C. and A.D. 900 approximately have not yet been identified, but they probably exist.

On the opposite extreme the situation is clearer; even though there are no dates associated to the metallurgy of the two last centuries of independent Cañari occupation, we know that it was a flourishing activity when the Incas arrived and that an interbreeding took place resulting in metallurgy with mixed characteristics between Cañari tradition and Inca state regulations. The proved sequence for Cañari metallurgy is, for the time being, between A.D. 900 and A.D.

Technology

An undetermined, though astonishing quantity of Cañari objects was lost during the lootings of Cañar and Azuay preventing us from studying the metallurgical techniques of that region. What we have nowadays gives us just a very limited view the processes, alloys and techniques handled in the past; this is why we cannot fully appreciate what this industry was really like. We shall start by studying the materials, metals and alloys that were known and used.

The most abundant metal was copper that, in many instances, was gilded by fusion or sheet plating. Silver was also employed; probably the most common form was the usual silver – copper alloy that is quite scarce in present day collections. With respect gold the analyses of Barrandon, Valdez y Estévez (2002) of four Cañari objects reveal the presence of silvery golds (69 to 82.2% Au with 6.6% to 27.4% Ag) with little copper (1.6 to 11.2% Cu); something that may be interpreted as the use of mine gold with natural inclusions of silver and copper. The authors conclude that the Cañari metal smiths, as almost all the Sierra smiths, used gold from the mines of the southern Sierra that have large amounts of Antimony (Barrandon, Valdez y Estévez 2002)

The metallurgic and metallographic analyses of Escalera y Barriuso (1978) of a group of objects from Ingapirca give us a better insight into the characteristics of the copper metallurgy of the region. The composition of the objects is quite uniform; copper is present in proportions from 78.6 to 99.1% and arsenic between 1 y 4% approximately; this means that arsenical copper (or bronze) was intentionally produced. There are three exceptions, two tupos with tin instead of arsenic (probably Inca) and an ornament with a high content of iron, possibly a late intrusion. The remaining elements, including iron, gold, silver and others appear as natural impurities in small quantities.

The metallographies revealed cast structures with a *"deficient metallurgical treatment"* (Escalera y Barriuso 1978) as well as deformed structures and twins, typical of cold working and annealing. The authors conclude that the raw material for all the objects found in the graves came from just one source, unknown up to date. On the other hand, the objects in the domestic contexts were made with materials from different provenances. Copper is not f the native type, it comes from the incompletely oxygenated refinement (smelting) of a copper mineral, that left many impurities in the metal and the casting did not reach temperatures high enough so as to distribute the metal components homogenously (Escalera y Barriuso 1978). All the objects, except the needles, were initially cast and then cold worked with alternate annealing.

Casting, as a primary manufacturing technique, had a limited use. It was mainly used to provide pre-forms for objects that were afterwards finished by hammering, it is not usual to find objects that were only cast. Lost wax casting was employed for small objects such as spear thrower hooks and casting in bivalve or composite moulds for large copper objects like axes.

On the other hand hammering is very frequent; so are assembling techniques that permitted making tri-dimensional or complex objects from sheets and wires. The later ones were shaped to make ear pendants (object 60-2-4279). The assembling methods include flanges or folds used to join together embossed gold discs to silver backings (objects 60-2-4257 and 4258), and hooks. In the works by Uhle (1922), Saville (1924) and Heuzey (1870) there are descriptions that allow us to infer the use of certain techniques. Uhle (1922) mentions welding and rivets to make assemblages; in the drawings of Saville (1924) there can be seen metallic feathers riveted to a crown.

There is a special finishing technique hitherto unrecorded that results, as the technique involving silver inlays in gold described for Puruha, in bi-colour objects. In this case, apparently, the following steps were involved: 1) A silver object was made either by casting or hammering; 2) The object was fusion gilded with a copper –gold alloy of a low melting point; 3) By means of mechanical working and polishing the adherence was improved, obtaining a certain degree of solid inter-diffusion; 4) At this stage the surface gold layer was refined by oxidising the copper; 5) Once the surface was homogeneous some zones of it were scraped until the gold layer was removed exposing again the silvered matrix. The result is a bi-colour surface with gold and silver areas. The technique that was used also in the Puruha metallurgy is analogous to zoned scraping on tombacs described by Plazas (1977/78) for the Carchi – Nariño Group.

Uhle (1922) describes *"...oval plaques in which gold bands half a centimetre wide alternate with similar ones of silver"*; a description that could correspond to either of the bi-colour techniques described for Puruha and Cañari. Gold, silver and copper were also combined in wooden staffs and spear throwers plated with embossed metal sheets. Repousse was use also for discs, breastplates and nose ornaments; in those there is evidence of work on both sides. Sheets were also engraved and occasionally embossing and engraving were combined. Spondylus beads were inlayed in gold discs for decoration (Pérez et al 1995).

When analysing Cañari technology we must keep in mind the chronological factor, something that we partially ignore. The reason for this is that the Inca invasion brought with it important technological changes. One of the most evident is the introduction of tin bronze that displaced arsenical bronze previously used, as can be seen in Ingapirca (Escalera y Barriuso 1978). This and other changes, gradually assimilated by the Cañari smiths, led to the formation of new hybrid Inca-Cañari style that dominated the industry throughout the 15th century. The classification and study of the diverse phases of the metallurgical history may reveal the nature and depth of those changes.

Typology and classification

The Cañari Group, possibly comprising two or more development phases, is probably one of the most rich and varied of the Sierra. However, the available sample gives us only a restricted view of its repertoire of forms and functions, as can be seen in the following table that summarises the information for the Cañari objects in the collection of the Ministry of Culture (Quito):

Function	Form - Representation	Frequency	Percentage
Shank	Simple – total	1	0.97
Rattle	Simple – total	1	0.97

Function	Form - Representation	Frequency	Percentage
Chisel	Simple	1	
	With heel with ear shaped extensions	1	
	Total chisels	2	1.94
Pendant	Simple	1	
	Shaped as a tumi	1	
	Total pendants	2	1.94
Bead	Simple	1	
	Globular	1	
	Total beads	2	1.94
Spear thrower hook	Simple - total	3	2.91
Axe	Simple	3	
	With straight heel	1	
	With strangled heel	1	
	With drilled heel	17	
	Total axes	22	21.36
Sheet	Simple - total	2	1.94
Nose ornament	Simple - total	4	3.88
Ear pendant	Spiral - total	1	0.97
Breastplate	Circular - total	51	49.51
Tweezers	Simple - total	1	0.97
Projectile point	Whistler - total	4	3.88
Head breaker	Simple	1	
	Star shaped	1	
	Total head breakers	2	1.94
Tumi	Simple – total	3	2.91
Tupo	Simple – total	2	1.94
Total		103	100.00

Evidently the simple does not provide data that can be adequately considered as representative of the Cañari Group. Probably the composition of this sample resulted more from the circumstances of its acquisition than from the real proportions of the finds in this region. The sample says nothing reliable about the proportions of the objects in the Group; it just indicates what types of objects exist. I this sense it is interesting to notice the presence of varied ornaments for the head, face and upper torso, as well as several instruments, tools and utensils. Several categories that we know about are missing in this collection.

Uhle (1924) gathered some information that, at his time, was still fresh about finds in Azuay and Cañar and mentioned among the objects found in graves (absent in the analysed sample) numerous staffs, spear throwers, textiles covered with metallic plaques, vases, large solid shanks, gold bands that enveloped the skeletons, crowns, bracelets, defensive shields, different types of flutes and staff heads with bird figures.

Saville (1924) also recorded complex crowns with ornaments and danglers, diadems, Tinculpa type breastplates, plates, embossed plaques, tubular ornaments sewn onto textiles, needles, weapons that combined head breaker and axe, zoomorphic figures and gold and silver feathers. Heuzey (1870) paid special attention to the elaborate spear throwers, In Ingapirca there were found hoops, rattles, axes and tumis (Fresco 1984). Mayer (1992) illustrates axes of many shapes from Cañar, including axe-monies and several "plaque emblems" similar to those of the Milagro – Quevedo Group. Jijon y Caamaño (1920) described special Tinculpa type pendants from this zone.

Figure 55 Cañari gold diadem: 6.9 x 9.7 x 1.3 cms.

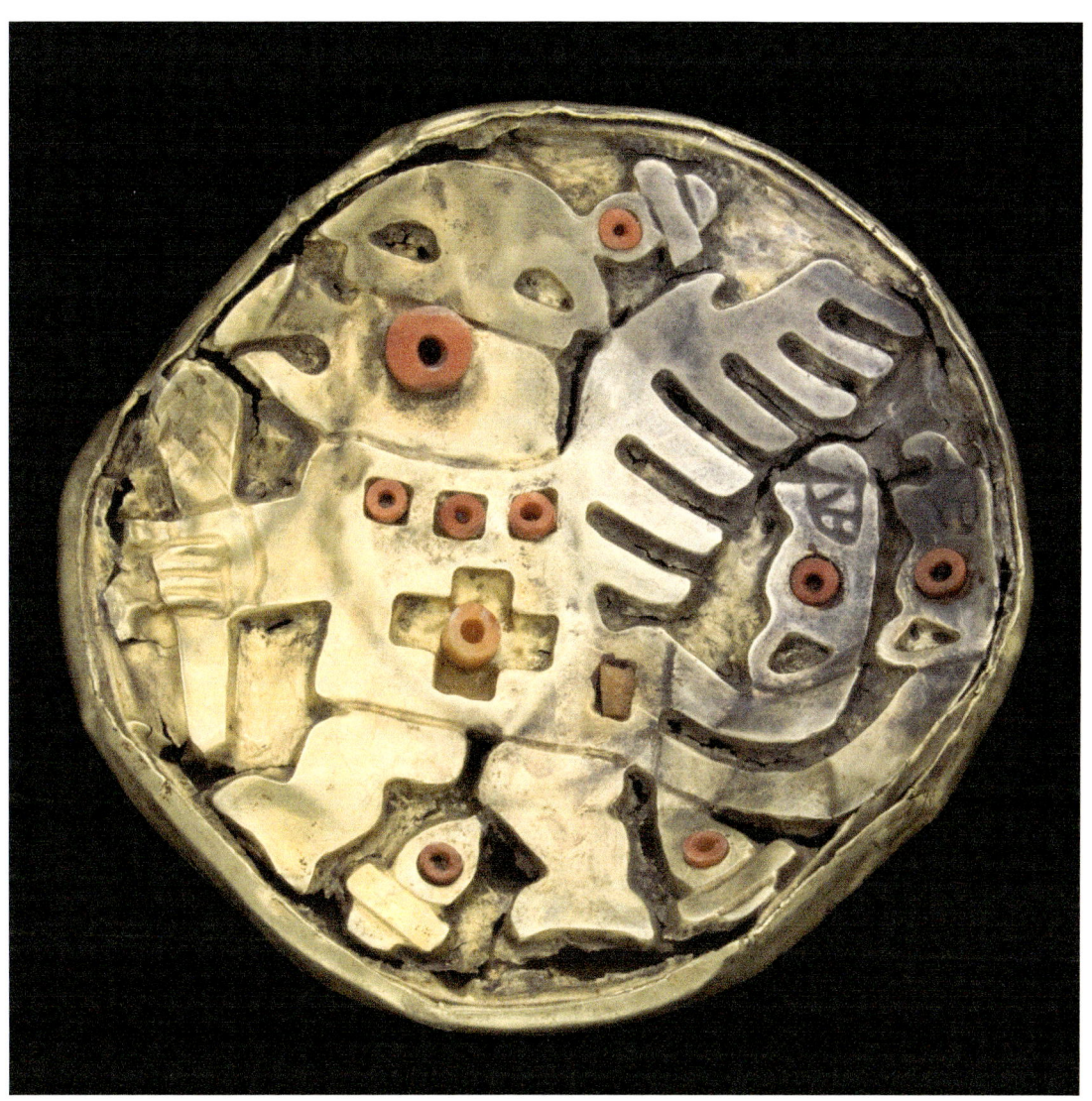

Figure 56 Cañari gold with Spondylus inlays ear pendant lid: 6.3 x 1 cms.

Figure 57 Cañari gold ear pendant lid: 8.8 x 0.3 cms.

Great Regional Groups: Cañari

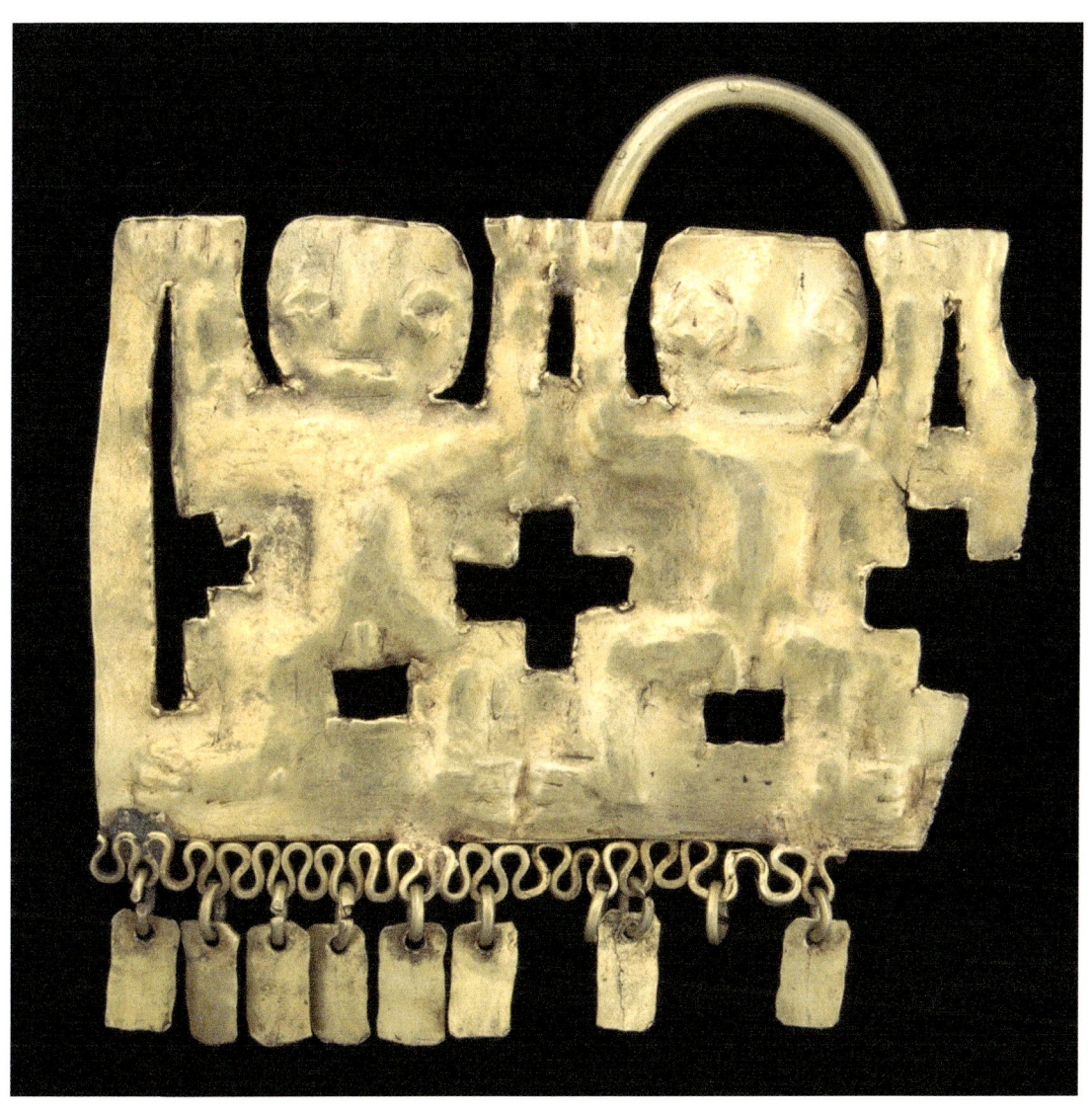

Figure 58 Cañari gold pendant with anthropomorphic figures: 7.4 x 6.9 x 0.4 cms.

Chapter 12

Great Regional Groups: Carchi – Nariño

The Carchi – Nariño Regional Group has been the subject of many investigations on both sides of the border between Colombia and Ecuador and it has, therefore, an important corpus of archaeological information. The interpretation of the archaeological components of the Group has varied greatly over the past forty years, but most of the recent research reports have had little diffusion.

Geographic Distribution

The collection of the Ministry of Culture corresponding to the Carchi – Nariño Group comprises a significant quantity of objects with the following spatial distribution:

Province	Canton	Number of objects	Percentage of total
Unknown	Unknown	19	3.9
Carchi	El Angel	1	0.2
	El Carmelo	121	24.5
	El Playón	9	1.8
	Huaca	28	5.7
	Montufar	10	2.0
	Carchi undefined	129	26.2
Imbabura	Atuntaqui	2	0.4
	Ibarra	5	1.0
	Pimampiro	147	29.8
Pichincha	Quito	22	4.5
	Total	493	100

In the following map the correlation between the distribution of the provenance sites for Carchi – Nariño objects (green dots) and all the other objects (red dots) in the collection of the Ministry of Culture is displayed (page opposite).

The distribution of Carchi – Nariño metallurgy is coherent with the rest of the archaeological information and with the limits of the territories of the northern Ecuador ethnic groups in the 16th century. The objects from Quito belong to the excavations of Doyon (2002) in the site of La Florida where various graves were found with funerary offerings composed, among other things, by gold objects that the author classified as part of the Chaupicruz Phase. This type of metallurgy is one of the initial styles of the Carchi – Nariño Group and is to be found later in the centre of the Nariño Department in Colombia (Gómez y Lleras 2006).

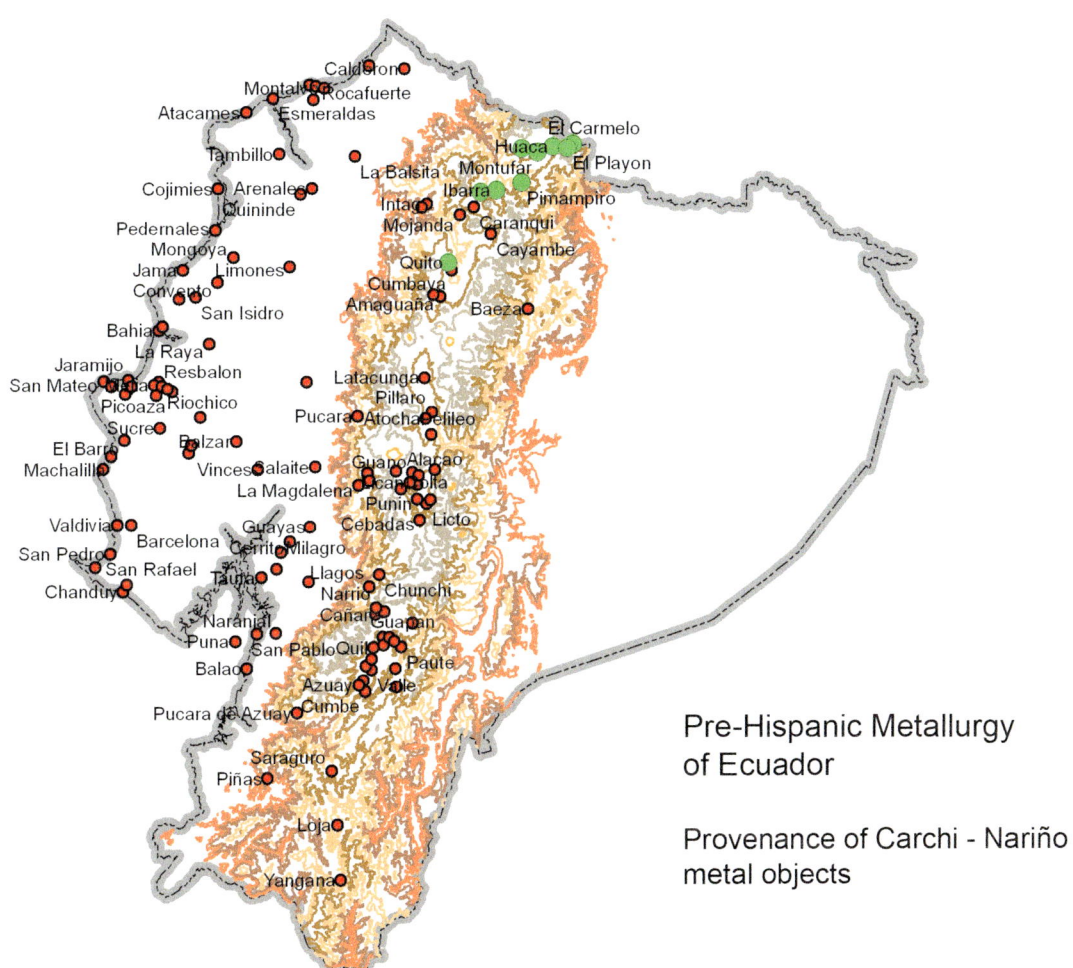

Pre-Hispanic Metallurgy of Ecuador

Provenance of Carchi - Nariño metal objects

In Carchi, apart from a noticeable quantity of objects without provenance information or with imprecise information (nearly 30%), there are two important foci of concentration; the first one is the Canton of El Carmelo (more than 24% of the objects), located on the Eastern slopes of the Cordillera, close to the Colombian border. There are no known reasons to explain why so many finds come from there, though we cannot discard that there was an important pre-Hispanic cemetery in the zone. On the other hand it might be also that objects found elsewhere could have been declared as found in this site. This is a common practice among treasure hunters.

The other site with an unusual quantity of finds is Pimampiro (30% approximately). The site was, at least in the 16th century and the immediately preceding decades, an important place of Exchange and cultivation; its location in the warm valley of the Chota River gave access to agricultural lands where coca could be grown, from there it was taken to the cold plateaux to the north and south of the valley. As a result it became an important regional power centre where the elites must have accumulated large quantities of metal objects.

This panorama would be incomplete if we were not to include the finds of the Andean region of the Department of Nariño and the south of the Department of Cauca (Colombia) that are in the *Museo del Oro* of Bogota. The following table summarises the finds:

Department	Municipality	Number of objects	Percentage of total
Unknown	Unknown	21	0.9
Cauca	Arboleda	1	0.04
	Cauca undefined	2	0.08
Nariño	Colon	1	0.04
	Consaca	92	3.8
	Córdoba	66	2.8
	Cumbal	2	0.08
	El Tambo	42	1.8
	Guachucal	50	2.1
	Guaitarilla	3	0.1
	Ipiales	243	10.1
	La Cruz	86	3.6
	La Unión	15	0.6
	Pasto	18	0.8
	Puerres	2	0.08
	Pupiales	778	32.5
	San Lorenzo	1	0.04
	Sandona	45	1.9
	Tangua	3	0.1
	Yacuanquer	34	1.4
	Nariño undefined	892	37.2
	Total	2,397	100

In the following map the graphic representation of the provenances of Carchi – Nariño objects in southern Colombia:

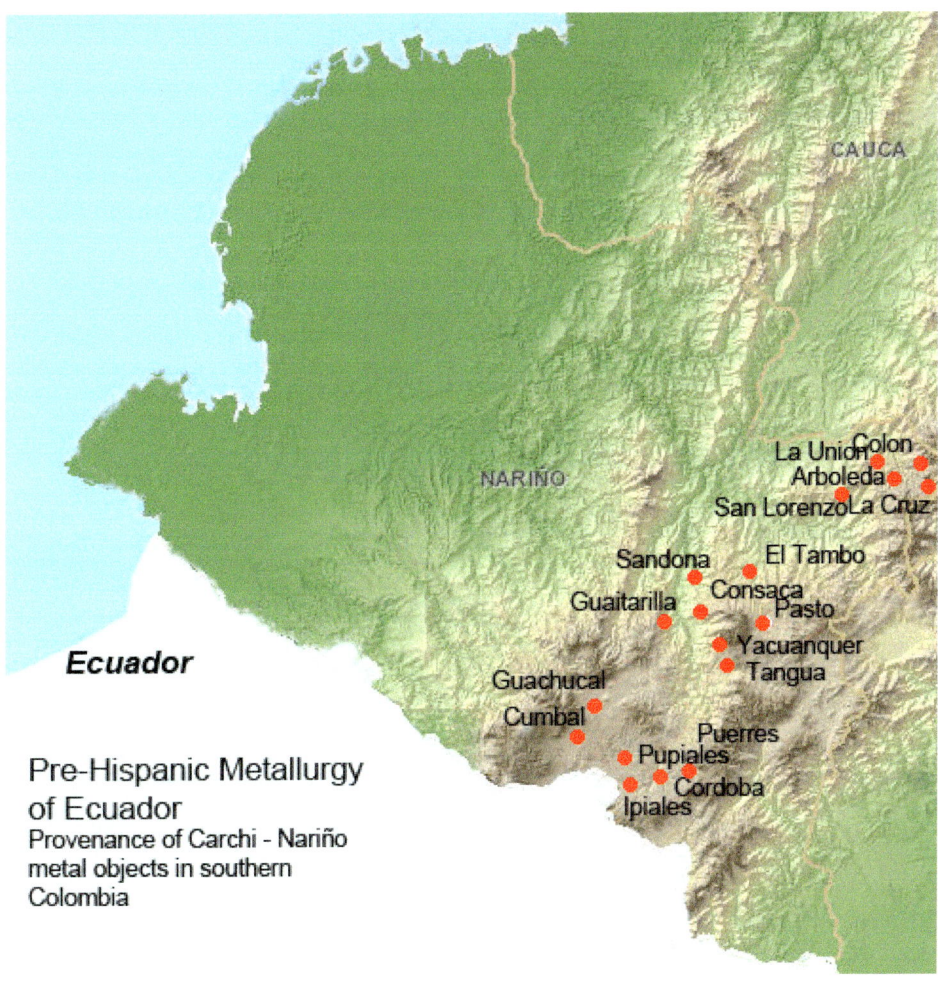

Pre-Hispanic Metallurgy of Ecuador
Provenance of Carchi - Nariño metal objects in southern Colombia

In this region the finds are distributed in an important number of sites located in different environments. There are no concentration foci with the exception of the municipalities near Ipiales, Pupiales and Guachucal where almost 45% of the objects were found. In this case the explanation comes from the finds of the legendary pre-Hispanic graveyards of Miraflores and Las Cruces where during many years hundreds of objects were looted. Apart from these sites there are minor concentrations in the north (La Cruz, La Unión and El Tambo), the central valley of the Guaitara River (Consaca, Sandona) and the south centre (Córdoba, Yacuanquer). The resulting pattern is an ample distribution with minor concentration foci spatially distant and a major concentration focus in the extreme south of the Department.

Chronology

This Group has a significant number of associated dates, thus allowing us to establish in detail its chronology. In the following table we summarise this information:

Date	Municipality/Canton	Laboratory Number	Style	Bibliographic Reference
A.D.130 ± 210	P-Quito	I-14968a	Yacuanquer	Doyon 1995
A.D.150 ± 40	N-Undefined	Beta-168164	Tinculpa	Museo del Oro 2002
A.D.331 ± 36	CR-El Carmelo	Beta-140431	Tinculpa	Museo del Oro 2002
A.D.340 ± 80	P-Quito	I-14969	Yacuanquer	Doyon 1995
A.D.350 ± 80	P-Quito	I-15348	Yacuanquer	Doyon 1995
A.D.400 ± 60	N-La Cruz	Beta-129140	La Cruz	Museo del Oro 1999
A.D.420 ± 80	P-Quito	I-14968b	Yacuanquer	Doyon 1995
A.D.440 ± 50	N-Yacuanquer	Beta-065196	Yacuanquer	Museo del Oro 1993
A.D.585 ± 65.	N-Yacuanquer	AA-12488	Yacuanquer	Cortés 1998
A.D.600 ± ?	P-Quito	No information	Undetermined	Molestina pers.com.
A.D.630 ± ?	P-Quito	No information	Undetermined	Molestina pers.com.
A.D.680 ± ?.	P-Quito	No information	Undetermined	Molestina pers.com.
A.D.810 ± 60	N-Pupiales	Beta-067955	Piartal	Plazas 1998
A.D.845 ± 80	N-Pupiales	IAN-50	Piartal	Plazas 1977-78
A.D.950 ± 40	N-Undefined	Beta-148002	Piartal	Museo del Oro 2001
A.D.950 ± 50	N-Ipiales	Beta-148001	Capuli	Museo del Oro 2001
A.D.1000 ± 40	N-Pupiales	Beta-144492	Piartal	Museo del Oro 2000
A.D.1050 ± 40	N-Pupiales	Beta-168165	Piartal	Museo del Oro 2002
A.D.1080 ± 115	N-Ipiales	IAN-67	Capuli	Uribe 1977-78
A.D.1120 ± 140	N-Pupiales	IAN-34	Piartal	Plazas 1977-78
A.D.1170 ± 40	N-Pupiales	Beta-144493	Capuli	Museo del Oro 2000
A.D.1250 ± 35	N-Pupiales	GRN-6911	Capuli	Cardale 1977-78
A.D.1290 ± 40	N-Pupiales	Beta-144490	Capuli	Museo del Oro 2000
A.D.1380 ± 50	N-Pupiales	Beta-270080	Capuli	Museo del Oro 2010
A.D.1470 ± 40	N-Ipiales	Beta-146384	Capuli	Museo del Oro 2000
A.D.1510 ± 30	N-La Cruz	Beta-133987	Late	Museo del Oro 1999
A.D.1600 ± 50	N-La Cruz	Beta-159602	Late	Museo del Oro 2001
A.D.1680 ± 50	N-La Cruz	Beta-159603	Late	Museo del Oro 2001

Taken altogether the Carchi – Nariño metallurgy has a development period of 1,600 years (A.D. 100 to A.D.1700 approximately); this is much longer than the chronological estimate of the Banco Central that goes from A.D. 700 to A.D. 1433 (Pérez et al 1995). However this approach is far too general; to properly understand the chronology of the Group it is necessary to examine each style independently. The research carried out on the *Museo del Oro* of Bogotá and the Banco Central del Ecuador (Ministry of Culture) (Gómez 2003, Gómez y Lleras 1999, 2006, Lleras y Gómez 2002, Lleras et al 2007) allowed us to re-evaluate the previous classifications (Plazas 1977/78) and to identify several different styles within the Regional Group whose chronological developments we shall discuss.

The Yacuanquer style, identified for the sites of La Florida, in Pichincha and Yacuanquer, in Nariño, is associated with dates that go from A.D. 130 to A.D. 585 this lasting approximately over 450 years. This chronology seems to confirm the existence of an early metallurgical tradition in the first centuries of the Common Era (Lleras et al 2007). A contemporary style, located in the north of Nariño, is La Cruz that, for the time being, is associated to one date of A.D. 400, thus providing only a preliminary chronology (Lleras et al 2007).

It seems that both styles were replaced in the region by new industries that were quite strong by the 9th century. The new style is the South-western Nariño – Carchi that comprises two sub-styles; the first one is Piartal, associated to dates that go from A.D. 810 to A.D. 1120, a period of just over 400 years (Lleras et al 2007). The second one, Capuli sub-style, is associated to dates that go from A.D. 950 to A.D.1470, thus lasting for over 500 years (Lleras et al 2007). Thus, the two sub-styles have slightly different chronologies, being somewhat earlier Piartal.

Another style, termed North Central Late, appears especially in the plateaux of northern Nariño (La Cruz) associated to dates that go from A.D. 1510 to A.D. 1680 and, consequently a duration of around 170 years (Lleras et al 2007).

The circular ear pendants, decorated in the middle with anthropomorphic, zoomorphic or anthropozoomorphic faces, termed Tinculpas (a Quechua Word) and a few objects related to them by shape or iconography (breastplates and pendants with one, two or three faces), have a special chronology of their own. They appear in association with objects from many of the major style and appear from the early phases up to the late ones (A.D. 130 to A.D. 1470) thus exhibiting a considerable temporal depth that conforms a true Tradition in the sense that this term has in archaeology (Lleras et al 2007).

Technology

There are over 300 composition analyses for objects of the Carchi – Nariño Group (Barrandon, Valdez y Estévez 2002, Rovira 1990 and Arqueometalurgia 2005) carried out by a wide range of analytical techniques (AAS, XRF, ICP-MS, OES, XRD, Microprobe and NAA); There are also more than 80 metallographies. It is impossible to make a detailed analysis of the information provided by these analyses, so we will limit ourselves to a synthesis of the general trends suggested by the data.

With regard to composition it can be said that copper rich matrices are predominant (tombacs with over 60%Cu) and there are many examples of almost pure copper (85% or more Cu), while gold-rich tombacs are scarce (60% or more Au) and much scarcer are gold matrices (85% or more Au). Silver is present in most samples analysed in proportions that range between 0.1% up to 88% Ag. In fact there are at least 14 objects that are silver – copper alloys (with little or no gold) and one that is a silver – gold alloy (75%Ag, 16%Au). There is platinum, as an impurity, in about 5% of the objects. Other minor elements present are nickel, antimony, zinc, iron, aluminium, arsenic and lead.

Metallographies reveal deformed and re-crystallised structures with twins due to cold working and annealing, as well as hot mechanical work (forge) at temperatures below the point of re-crystallisation. In some cases hammering was the last step, in others annealing was last. There are a few samples of cast structures and possible evidences of autogenous welding. The existence of fusion and depletion gilding and depletion silvering plating was also confirmed.

At this point it is important to note that the study of Carchi – Nariño metallurgic technology has to take into account the division of the Group into its different styles and sub-styles. Apart from basic techniques, casting and hammering, there are much more technological traits that differentiate these styles than those that assimilate them. Let us consider in the first place the Yacuanquer style, probably the earliest in the region.

In this style no cast objects have been recorded, all the pieces were made by hammering in copper rich tombacs and fusion gilded (Gómez y Lleras 2006); there are also some gold matrices (Doyon 2002). Fusion gilding was known in several areas of Ecuador, the technique was described by Scott (1985) and confirmed for these objects by Cortés (1997) based on metallographies and SEM and EDS analyses that show that the gold layer was applied only on one of the surfaces of the objects.

Garzón et al (2007) found traces of silver, gold, platinum and arsenic in some objects; the first three metals are associated to the use of alluvial gold from the Pacific coast for the platings while arsenic is a natural impurity of the copper mineral used to make the objects. The gilding layer is not pure gold but a rich gold tombac that was poured over a sheet of a copper rich material that had been previously hammered and annealed; subsequent cold working and cleaning enriched the surface of the layer on the side that was not fusion gilded. Hammering was deficiently controlled because there is a certain harshness of the metal that was not thoroughly eliminated by annealing (Garzón et al 2007).

The only other observations with respect to the technology of this style are referred to the assemblage of sheets by means of hooks and the repairs made with rivets that allowed the use of the objects after they had suffered fractures. The high degree of corrosion of the objects makes it difficult to assess the degree of polishing or the characteristics of decoration, should it exist.

The second style is La Cruz Early that, up to date, has been found only in the north of the Department of Nariño, Colombia. The objects of this style were made by hammering, the matrix material is a copper rich alloy that was worked mechanically with various cycles of annealing; this treatment avoided the harshness of the metal whose microstructure revealed polygonal grains and deformed annealing twins (Garzón et al 2007). One of the surfaces of the objects was fusion gilded with a gold rich tombac with a lower melting point than the matrix; then the object was cold worked favouring inter-diffusion between the gilding layer and the matrix and enriching its surface by depletion. The other surface was gilded by depletion, probably in an accidental way while working the front surface (Garzón et al 2007).

In this style there are also examples of the assemblage of sheets with hooks, probably to make loincloths (Gómez y Lleras 2006). Embossing as a decoration technique is present in several objects; both sides were worked. Good polishing can be observed in the surfaces that have resisted corrosion.

The North Central Late style of Nariño is also manufactured almost exclusively by hammering gold and tombac matrices that were gilded by depletion. In technological terms it is a simple group mainly composed of small ornaments with little decoration except occasional embossing. Breastplates exhibit a well-controlled mechanical work and excellent surface polishing. God wire was made and shaped to form adornments.

The style termed South-western Nariño – Carchi (Gómez y Lleras 1999, 2006, Lleras et al 2007) is divided, in turn, into two sub-styles. The Capuli sub-style is characterised technologically by the following characteristics: gold is predominant, though depletion gilded tombac as well as silver were also used (Gómez y Lleras 2006). Hammering was the preferred technique but, for certain small objects, lost wax casting was employed. Hammering used most times cast pre-forms.

There are some objects assembled with hooks and rivets. However, the preferred technique for assemblage was welding, both the autogenous and the type that uses auxiliary alloys and fluxes and even granulation welding. The natural complements of hammering, cutting and embossing were used for decorative purposes in order to make complex openwork and to enhance designs, working on both sides of the object. Finishing is uneven; some objects show a deficient polishing that reveals the marks of hammering while others have excellent polishing (Gómez y Lleras 2006).

The second sub-style, Piartal, is characterised by the following technological traits: hammering is so predominant that there are only very scarce examples of lost wax casting. Gold – copper alloys (tombacs) were used but usually they have such a high natural silver content that they become real ternary alloys (Au-Cu-Ag). As a general rule these tombacs have much more copper than gold but they have yellowish surfaces due to depletion gilding, the only gilding technique known for this sub-style (Gómez y Lleras 2006). Danglers were assembled to the major object by means of hooks; the components of musical instruments (flutes) were tied together with wire.

The most relevant trait from the technological point of view in this style is a technique of surface treatment unique in the Andean region. The technique is known as zoned scraping and was initially described by Plazas (1977-8). The starting point is a tombac matrix that is enriched by depletion gilding and then carefully polished (in a radial direction in the case of discs and circular breastplates) and afterwards burnished. Next, one of two paths can be followed, or both in sequence. The first alternative is to scrape part of the enriched surface following the desired design; in this way a combination of colours is obtained, gold yellow for the enriched surface and tombac pink for the scraped surfaces. The other alternative is to protect, with wax or resin, part of the surface following the desired design; then the unprotected surface is attacked with an acid or other corrosive agent; the result is a combination of textures, brilliant in the protected zone and mate in the attacked zone (Plazas 1977-8). If both alternatives are applied the result is an object with two colours and two textures, an unusual visual play obtained without the use of platings or inlays.

The technological panorama of the Carchi – Nariño Group is relatively heterogeneous. Nevertheless, there are general trends. The use of copper and copper rich tombacs is predominant; the objects are generally worked by hammering and gilded by fusion or depletion. There is also some embossing and assemblage with emphasis on welding. Less important materials and processes include silver, pure gold, lost wax casting and assemblage with hooks.

Typology and classification

For the purpose of analysing form and function of the Carchi – Nariño Regional Group it is desirable to be able to separate the collections into the above mentioned styles, because each one has different form and function characteristics. This will be done, as far as possible, taking into account the restrictions due to the state of the classificatory systems of the collections of the Ministry of Culture (Ecuador) and *Museo del Oro* (Colombia). Neither collection is completely classified according to the criteria of the styles; there are in both collections a large number of objects that are provisionally classified as Carchi – Nariño undetermined.

The most detailed presentation that is possible, for the time being, involves a first table dedicated to the Early Period in which the objects of the Yacuanquer and Early La Cruz styles for both collections are included. The second table is dedicated to the objects of the South-Western Nariño – Carchi style (with its two sub-styles, Capuli and Piartal) in the collection of the Ministry of Culture. The third table corresponds to the same style for the collection of the *Museo del Oro*. The fourth table includes the objects of the North – Central Late style for both collections. Tables five and six are for the objects classified as Carchi – Nariño Undefined in the collections of the Ministry of Culture and the *Museo del Oro* respectively. The decision to join or not the objects of the two collections was taken mainly attending the frequencies; when those were too low in either collection it was not considered convenient to present a separate table, when those were high the collections were presented separately. The following table contains the objects of the two earliest styles of the region (Yacuanquer and Early La Cruz) in the collections of the Ministry of Culture (Quito) and *Museo del Oro* (Bogotá):

Function	Form - Representation	Frequency	Percentage
Pendant	Heart shaped	1	5.88
Crown	Simple	1	
	Truncated cone	1	
	Rectangular	1	
	Shaped as a band	1	
	Total crowns	4	23.53
Diadem	Simple	1	
	Composite shape	1	
	Total diadems	2	11.76
Breastplate	Square	1	
	Composite shape	1	
	Total breastplates	2	11.76
Container/bowl	Shaped as a gourd - total	6	35.29
Loincloth	Rectangular - total	2	11.76
Total		17	100.00

The reduced size of the sample makes it useless to attempt any type of statistical analysis. Apart from the body ornaments included in the table we can note that in the Yacuanquer style there are tumis, ear pendants and bracelets (Gómez y Lleras 2006) that are not recorded as such in the inventory. We still know very little about these early styles; the scarce data on form and function indicate an emphasis on body decoration, especially with large objects (crowns, diadems, breastplates) and on textile decoration de textiles. It is quite striking not to find some of the most common types of adornments like nose ornaments and bead necklaces. The objects recovered up to date probably come from the graves of important people within the social group; it is possible that many simple pieces that usually appear in other contexts (ear pendants, nose ornaments) do exist in these styles.

In the following table the objects of the South – Western style in the collection of the Ministry of Culture:

Function	Form - Representation	Frequency	Percentage
Textile applique	Circular	9	
	Triangular	1	
	Total textile appliques	10	10.87
Rattle	Globular	6	
	Semi-globular	1	
	Total rattles	7	7.61
Pendant	Composite shape schematic	1	
	Zoomorphic, shaped as a monkey	1	
	Shaped as a bird	4	
	Triangular with zoomorphic figure	1	
	Shaped as a flower	1	
	Of a major object	1	
	Total pendants	9	9.78
Ear pendant	Circular	20	
	Shaped as a bow	16	
	Circular with zoomorphic figure	2	
	Shaped as a bow with ornament	2	
	Oval	2	
	Total ear pendants	42	45.65
Bead necklace	Oval - total	1	1.09
Flute	Simple - total	1	1.09
Nose ornament	Semi-elliptical with lateral extensions	4	
	Semi-elliptical	3	
	Annular with zoomorphic figure	1	
	Total nose ornaments	8	8.70
Ear ornament	Annular with zoomorphic figure - total	2	2.17
Breastplate	Circular with rim	2	
	Elliptical	1	
	Shaped as a bird schematic	2	
	Circular	1	
	Total breastplates	6	6.52
Plaque	Circular - total	1	1.09
Plaque pendant	Trapezoidal - total	5	5.43
Total		92	100.00

In the following table the objects of the collection of the *Museo del Oro* belonging to the South-Western Nariño – Carchi style:

Great Regional Groups: Carchi – Nariño

Function	Form - Representation	Frequency	Percentage
Applique	Trapezoidal	3	
	Circular concave	1	
	Circular	1	
	Total appliques	5	0.81
Textile applique	Zoomorphic shaped as a bird	4	
	Circular	39	
	Trapezoidal	16	
	Composite shape	4	
	Circular star shaped	2	
	Star shaped	3	
	Zoomorphic shaped as a monkey	1	
	Circular with zoomorphic figure	2	
	Shaped as a fang	4	
	Triangular	2	
	Rhomboidal	1	
	Triangular with dangler	3	
	Composite shape with dangler	3	
	Circular convex	5	
	Circular with dangler	1	
	Anthropomorphic schematic	1	
	Rectangular	3	
	Square	1	
	Circular shaped as a flower	1	
	Elliptical	1	
	Total textile appliques	124	20.10
Bracelet	Cylindrical - total	2	0.32
Basket	Simple - total	2	0.32
Rattle	Semi-globular	74	
	Oval	1	
	Total rattles	75	12.16
Pendant	Star shaped	1	
	Zoomorphic shaped as a monkey	5	
	Zoomorphic shaped as a bird	13	
	Circular with dangler	1	
	Zoomorphic	1	
	Zoomorphic schematic bird	1	
	Circular	1	
	Anthropozoomorphic	1	
	Total pendants	24	3.89

Function	Form - Representation	Frequency	Percentage
Ear pendant	Circular	26	
	Circular convex	20	
	Shaped as a bow with zoomorphic figure	23	
	Circular with zoomorphic figure	6	
	Shaped as a bow with ornament	7	
	Shaped as a bow	4	
	Circular concave	30	
	Circular with dangler	1	
	Circular with anthropomorphic figure	3	
	Total ear pendants	120	19.45
Diadem	Simple	2	
	Imitating feathers	28	
	Total diadems	30	4.86
Spinning disc	Simple - total	23	3.73
Figure	Zoomorphic shaped as a monkey - total	5	0.81
Nose ornaments	Semi-elliptical with horizontal extensions	34	
	Semi-elliptical	114	
	Elliptical	1	
	Semi-lunar	9	
	Semi-lunar with ornament	2	
	Semi-lunar with zoomorphic figure	3	
	Semi-elliptical star shaped	1	
	Semi-elliptical convex	1	
	Total nose ornaments	165	26.74
Ear ornament	Annular with zoomorphic figure	4	
	Annular hollow	1	
	Total ear ornaments	5	0.81
Lime-dipper pin	With zoomorphic figure - total	1	0.16
Breastplate	Circular	13	
	Circular concave	6	
	Circular convex	5	
	Circular with geometric figure	1	
	Elliptical	2	
	Zoomorphic schematic bird	2	
	Circular with anthropomorphic and zoomorphic figures	4	
	Total breastplates	33	5.35
Dangler	Circular	1	
	Elliptical	1	
	Shaped as a claw	1	
	Total danglers	3	0.49
Total		617	100.00

In general terms the relative proportions of the objects in both collections are similar. There is a marked emphasis on the head and face (ear pendants, skin appliques, diadems, nose ornaments and ear ornaments) that account for 56.52% in the collection of the Ministry of Culture and 52.67% in the *Museo del Oro*. Other body ornaments are made to be worn on the chest (bracelets, pendants, bead necklaces, beads and breastplates) that account for 17.39% and 9.24% respectively. The remaining adornments (textile appliques, rattles, danglers) do not have a precise body location, except for bracelets, but they represent a significant proportion in both collections (23.91% and 32.75%).

Utensils and instruments (flutes, spinning discs, baskets, lime dipper pins) are scarce (1.09% and 4.21%) and unevenly distributed in both collections; it is striking, for example, that there are no spinning discs in the collection of the Ministry of Culture, since those are one of the most prominent objects in the Piartal sub-style. The South-Western Nariño – Carchi Group is strongly oriented towards the adornment of the face and head with less emphasis on the chest and clothes (textiles). No tools have been recorded and utensils are scarce, but those present have been carefully made. The general conclusion derived from the analysis of form and function is that this is a style aimed at elite individuals, something that agrees with the conclusions of previous works in this region (Gómez 2003; Gómez y Lleras 1999, 2006).

In the following table we have included the objects of the North Central Late style in the collections of the Ministry of Culture (Quito) and *Museo del Oro* (Bogotá):

Function	Form - Representation	Frequency	Percentage
Ear pendant	Circular with geometric figure	1	
	Circular with zoomorphic figure	2	
	Total ear pendants	3	3.95
Diadem	Shaped as an H, anthropomorphic figure	1	
	Shaped as an H	5	
	Total diadems	6	7.89
Nose ornament	Semi-lunar with spiral ornament	10	
	Annular with spiral ornament	1	
	Semi-lunar with zoomorphic figure	1	
	Annular with ornament	1	
	Semi-lunar	34	
	Total nose ornaments	47	61.84
Ear ornament	Annular with spiral ornament	11	
	Semilunar with ornament	2	
	Semilunar with spiral ornament	6	
	Total ear ornaments	19	25.00
Container/bowl	Shaped as a gourd	1	1.32
Total		76	100.00

We are before another case in which the limited size of the simple severely restricts statistical analysis. However, it is extremely interesting to note that almost all the objects (98.68%) in this style are made to be worn on the face and head. Most objects are simple and small, except for breastplates (Gómez y Lleras 2006). Apart from the objects listed only springs have been reported (Gómez y Lleras 2006); those and breastplates have not yet been recorded in the inventory of this style. This style, though mainly decorative and lacking utensils and tools, seems to be aimed at an extended use by the majority of the population and not by elite people. Probably this is why there are some many nose ornaments and ear pendants, small and simple and no large complex objects. In this sense there would have been a significant change of orientation of the metallurgic production.

In the following table we have included the objects of the Carchi – Nariño Group in the collection of the Ministry of Culture that have not been assigned to any particular style:

Function	Form - Representation	Frequency	Percentage
Sub-labial Adornment	Anthropomorphic - total	1	0.30
Wire	Spiral shape- total	1	0.30
Staff shank	Annular	2	
	Cylindrical	18	
	Total staff shanks	20	5.93
Applique	Circular with zoomorphic figure	1	
	Circular concave, zoomorphic figure	1	
	Total appliques	2	0.60
Textile applique	Circular - total	1	0.30
Shank	With dangler - total	2	0.60
Bracelet	Band shaped - total	1	0.30
Rattle	Simple	2	
	Globular with zoomorphic figure	1	
	Total rattles	3	0.89
Chisel	Simple - total	1	0.30
Pendant	Simple	9	
	Of a major object	1	
	Composite shape	1	
	Globular zoomorphic	2	
	Double	1	
	Circular, anthropozoomorphic figure	1	
	Semi-lunar	3	
	Shaped as a bird	3	
	Circular	3	
	Cylindrical with zoomorphic figure	1	
	Total pendants	25	7.42

Function	Form - Representation	Frequency	Percentage
Ear pendant	Shaped as a bow	1	
	Circular	62	
	Circular Tinculpa type	1	
	Circular concave	4	
	Total ear pendants	68	20.18
Bead necklace	Simple	3	
	Various shapes, dangler	5	
	Oval	1	
	Anthropomorphic schematic	1	
	Cylindrical	12	
	With dangler	1	
	Globular	4	
	Globular anthropomorphic schematic	1	
	Cylindrical anthropomorphic schematic	1	
	Various shapes	3	
	Total bead necklaces	32	9.50
Bead cover	Semi-globular - total	1	0.30
Bead	Simple	2	
	Rectangular	1	
	Shaped as a gourd	6	
	Anthropomorphic schematic	8	
	Cylindrical, lateral extensions	2	
	Trapezoidal	1	
	Bi-conical	4	
	Total beads	24	7.12
Diadem	Rectangular	2	
	Rectangular band shaped	1	
	Band shaped	42	
	Total diadems	45	13.35
Sheet	Simple	15	
	Double	1	
	Elongated	2	
	Total sheets	18	5.34
Nose ornament	Simple	2	
	Shaped as a N	2	
	Semi-lunar with adornment	1	
	Semi-lunar	4	
	Circular	1	
	Total nose ornaments	10	2.97

Function	Form - Representation	Frequency	Percentage
Ear ornament	Shaped as a fork, circular ending	19	
	Zoomorphic	2	
	Bird shaped	2	
	Semi-lunar	2	
	Shaped as a feline	2	
	Shaped as a fork	2	
	Total ear ornaments	29	8.61
Breastplate	Simple	9	
	Zoomorphic	2	
	With shell	1	
	Circular	10	
	Shaped as a head anthropomorphic	1	
	With zoomorphic figure	2	
	Shaped as a head	2	
	Shaped as a head zoomorphic	1	
	Total breastplates	28	8.31
Nipple cover	Circular concave	2	
	Circular convex	2	
	Spiral shaped	1	
	Conical	10	
	Total nipple covers	15	4.45
Plaque	Circular - total	2	0.60
Dangler	Rectangular	2	
	Circular	1	
	Triangular	1	
	Total danglers	4	1.19
Spring	Simple - total	1	0.30
Lid	Circular - total	1	0.30
Tumi	Simple - total	1	0.30
Tupo	Shaped as a feline, rattle - total	1	0.30
Total		337	100.00

In the following table we included the objects of the Carchi – Nariño Group in the *Museo del Oro* that have not been assigned to any particular style:

Function	Form - Representation	Frequency	Percentage
Pin	With semi-circular ending - total	1	0.06
Fishing hook	Simple - total	1	0.06
Skin applique	Semi-globular - total	1	0.06

Function	Form - Representation	Frequency	Percentage
Textile applique	Composite shape	2	
	Circular	3	
	Shaped as a hook	2	
	Circular convex	1	
	Circular concave	1	
	Star shaped	1	
	Total textile appliques	10	0.64
Ingot	Simple - total	2	0.13
Bracelet	Annular	10	
	Cylindrical	1	
	Cylindrical open	2	
	Total bracelets	13	0.84
Bell	Conical - total	1	0.06
Rattle	Globular	2	
	Elliptical	1	
	Semi-globular	5	
	Conical	2	
	Total rattles	10	0.64
Chisel	Simple - total	9	0.58
Pendant	Zoomorphic shaped as a bird	5	
	Zoomorphic shaped as a frog	1	
	With dangler	1	
	Circular	2	
	Zoomorphic shaped as a snail	1	
	With dangler	1	
	Shaped as a fang	2	
	Circular with zoomorphic figure	2	
	Square	2	
	Total pendants	17	1.10
Ear pendant	Circular	276	
	Circular convex	7	
	Circular with zoomorphic figure	2	
	Square	1	
	Elliptical	2	
	Circular concave	3	
	Circular with anthropomorphic figure	2	
	Circular with geometric figure	1	
	Total ear pendants	294	18.94
Necklace	Annular - total	1	0.06

Function	Form - Representation	Frequency	Percentage
Bead necklaces	Bi-conical	2	
	Oval	1	
	Various shapes	7	
	Elliptical	2	
	Circular	2	
	Cylindrical	6	
	Globular	5	
	Total bead necklaces	25	1.61
Crown	Simple - total	2	0.13
Staff cover	Simple	4	
	With anthropomorphic figure	1	
	Total staff covers	5	0.32
Flute cover	Simple - total	1	0.06
Bead	Bi-conical	2	
	Cylindrical rattle	1	
	Cylindrical	2	
	Cylindrical with zoomorphic figure	1	
	Globular	9	
	Cylindrical hollow	2	
	Cylindrical with dangler	1	
	Total beads	18	1.16
Diadem	Imitating feathers	1	
	Shaped as an H	1	
	Shaped as an H with anthropomorphic figure	1	
	Total diadems	3	0.19
Figure	Anthropomorphic of a major object - total	1	0.06
Pan flute	Simple - total	5	0.32
Axe	Simple	1	
	Trapezoidal	9	
	Shaped as an anchor	2	
	Total axes	12	0.77
Tool	Triangular - total	1	0.06
Sheet	Rectangular - total	3	0.19

Great Regional Groups: Carchi – Nariño

Function	Form - Representation	Frequency	Percentage
Nose ornament	Twisted, horizontal extensions	2	
	Horizontal extensions, ending	1	
	Semi-elliptical convex	223	
	Semi-elliptical	158	
	Composite shape	1	
	Semi-lunar	390	
	Twisted	1	
	With ascending extensions	1	
	Annular	22	
	Semi-lunar with spiral ornaments	2	
	Shaped as a N with endings	1	
	Total nose ornaments	802	51.68
Ear pendant	Annular	151	
	With dangler	1	
	Fork shaped, zoomorphic figure	1	
	Fork shaped	6	
	Fork shaped, circular ending	7	
	Fork shaped, dangler	2	
	Annular with spiral ornament	3	
	Annular with ending	15	
	Shaped as a flower with zoomorphic figure	3	
	Annular hollow	10	
	Fork shaped, spiral ornament	2	
	Total ear pendants	201	12.95
Breastplates	Rectangular with lateral flap	1	
	Circular	16	
	Zoomorphic shaped as a bird	1	
	Circular with geometric figure	1	
	Circular with anthropomorphic figure	1	
	Elliptical with anthropomorphic figure	2	
	Rectangular	2	
	Circular with zoomorphic figure	1	
	Elliptical	3	
	Heart shaped	1	
	Total breastplates	29	1.87
Nipple cover	Conical	27	
	Truncated cone	4	
	Total nipple covers	31	2.00
Tweezers	Trapezoidal	1	
	Globular with extension	1	
	Total tweezers	2	0.13
Plaque	Circular - total	7	0.45

Function	Form - Representation	Frequency	Percentage
Dangler	Circular convex	1	
	Circular	1	
	Total danglers	2	0.13
Container/bowl	Shaped as a gourd	5	0.32
Spring	Simple - total	37	2.38
Total		1,552	100.00

The large number and variety of objects in the two last tables could, no doubt, complete our understanding of the styles of the Carchi - Nariño Group if they were properly classified. In the present state of knowledge this material can only point to the fact that there was a wide variation and a high frequency of body ornaments in this area during the long time of development of all the styles.

The great importance of head and face adornments can be confirmed; sub-labial adornments, skin appliques, ear pendants, crowns, diadems, nose ornaments and ear ornaments account for 46.14% of the collection of the Ministry of Culture and 83.95% of the collection of the *Museo del Oro*. In a second place we find adornments for the chest (necklaces, beads, pendants, breastplates, nipple covers and tupos) with 37.2% and 86% respectively. Adornments for other parts of the body (arms, belly, legs, etc.) are scarce. The group of tools, instruments and utensils is small in quantitative terms but it is interesting to note that new categories appear, such as chisels, springs, tumis, fishing hooks, ingots, axes, tools, flutes, bowls, etc.

Great Regional Groups: Carchi – Nariño

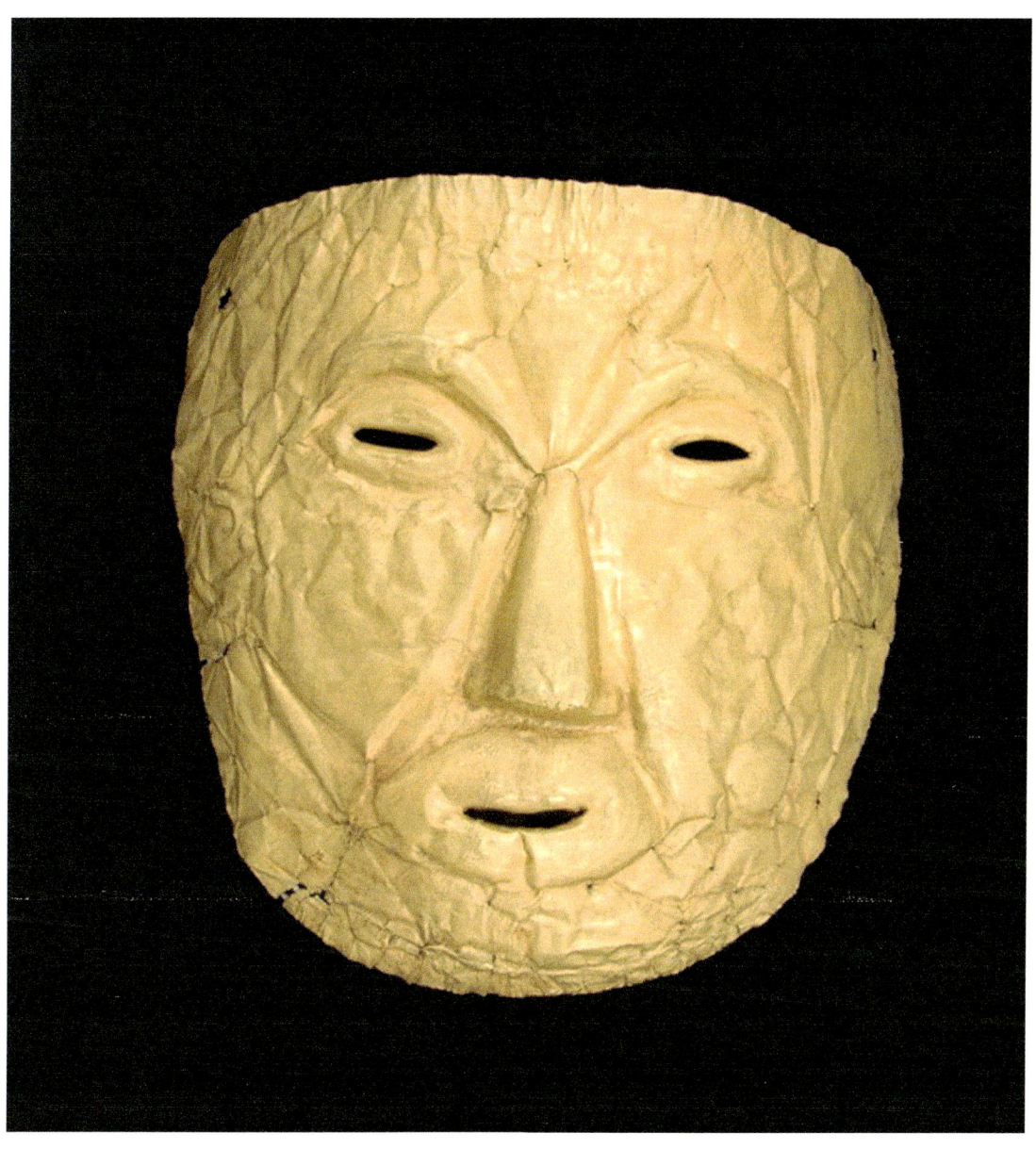

Figure 61 Carchi - Nariño gold mask: 14.6 x 9.4 x 1.8 cms.

160 Metallurgy in Ancient Ecuador

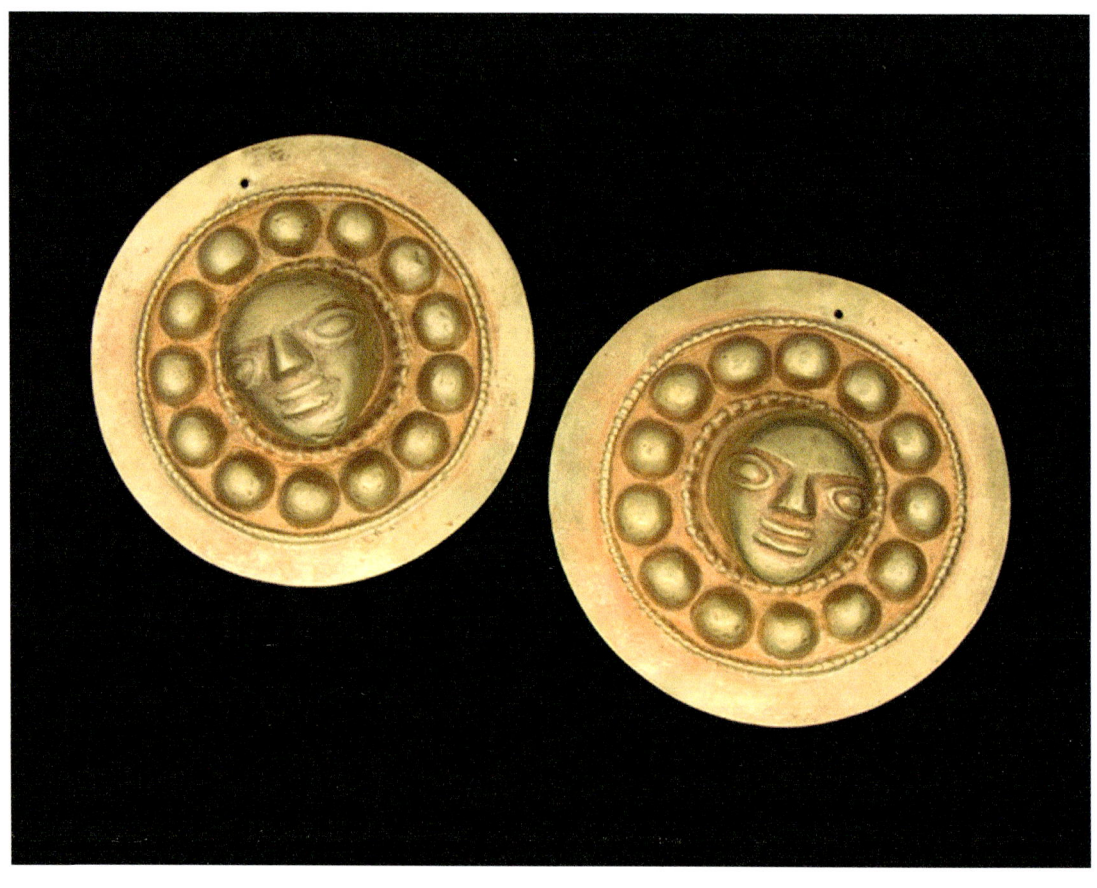

Figure 62 Carchi - Nariño tombac ear pendants, tinculpa style:
11.9 x 4 and 11.9 x 4 cms.

Great Regional Groups: Carchi – Nariño

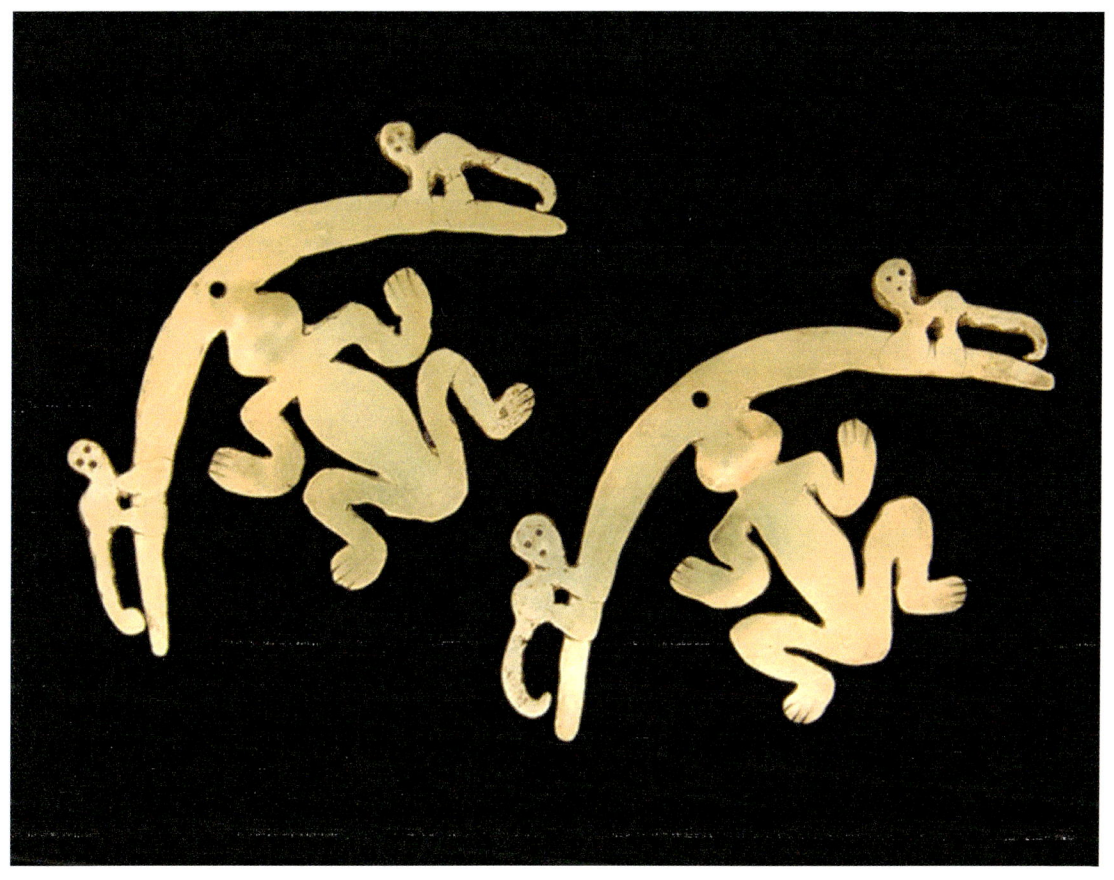

Figure 63 Carchi - Nariño gold ear pendants with zoomorphic figures:
6.6 x 10 x 0.4 and 7 x 9.8 x 0.4 cms.

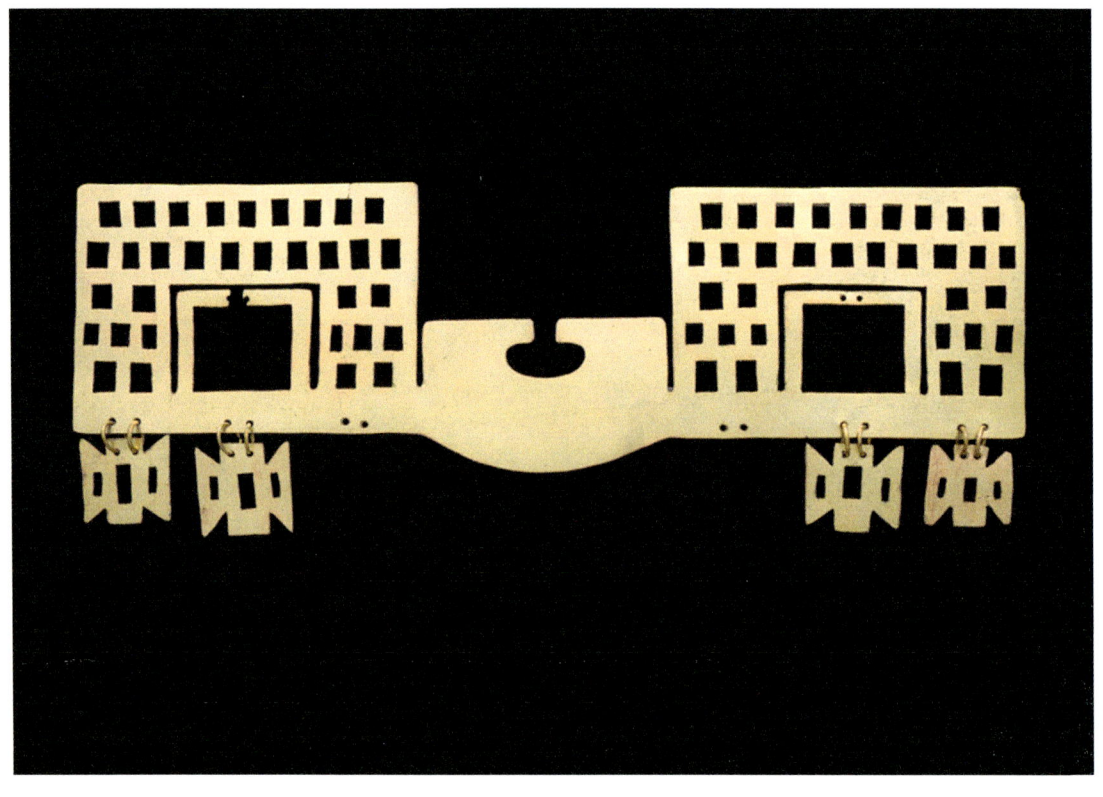

FIGURE 64 CARCHI - NARIÑO TOMBAC NOSE ORNAMENT: 6.5 X 16 X 0.5 CMS.

Great Regional Groups: Carchi – Nariño

Figure 65 Carchi - Nariño tombac necklace: 11 x 5 x 0.5 cms.

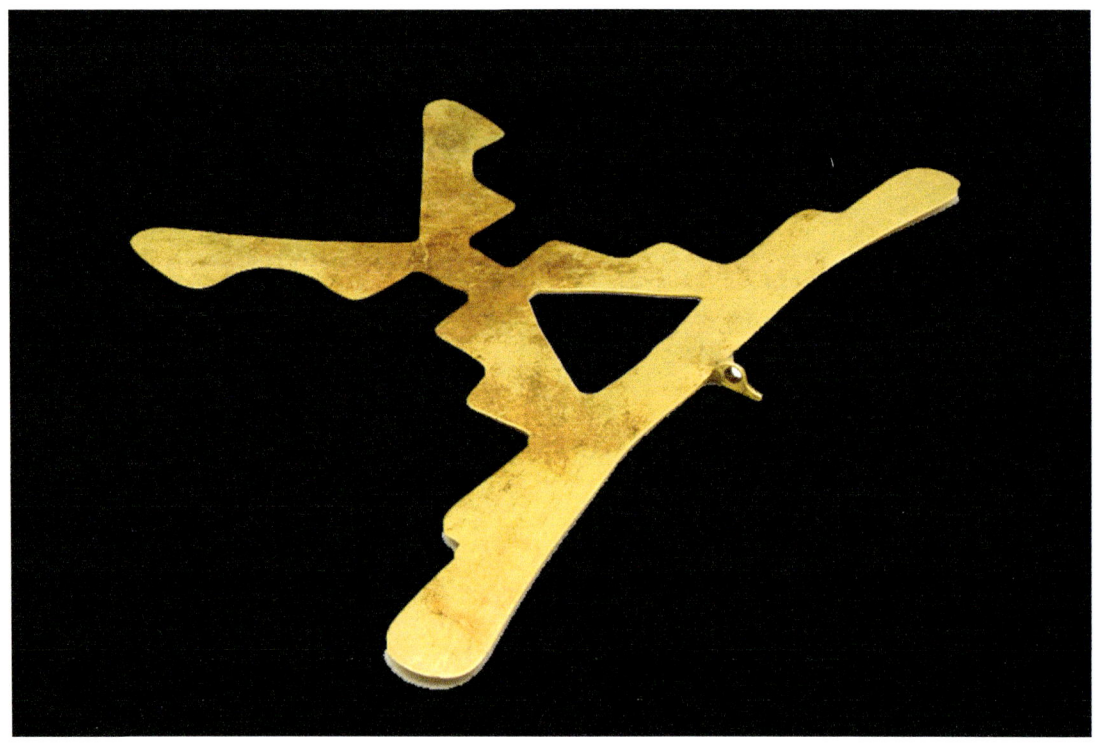

Figure 66 Carchi - Nariño gold pendant shaped as a bird: 13.3 x 16.8 x 0.5 cms.

Great Regional Groups: Carchi – Nariño

Figure 67 Carchi - Nariño gold pendant shaped as a bird: 13.2 x 11.4 x 1.8 cms.

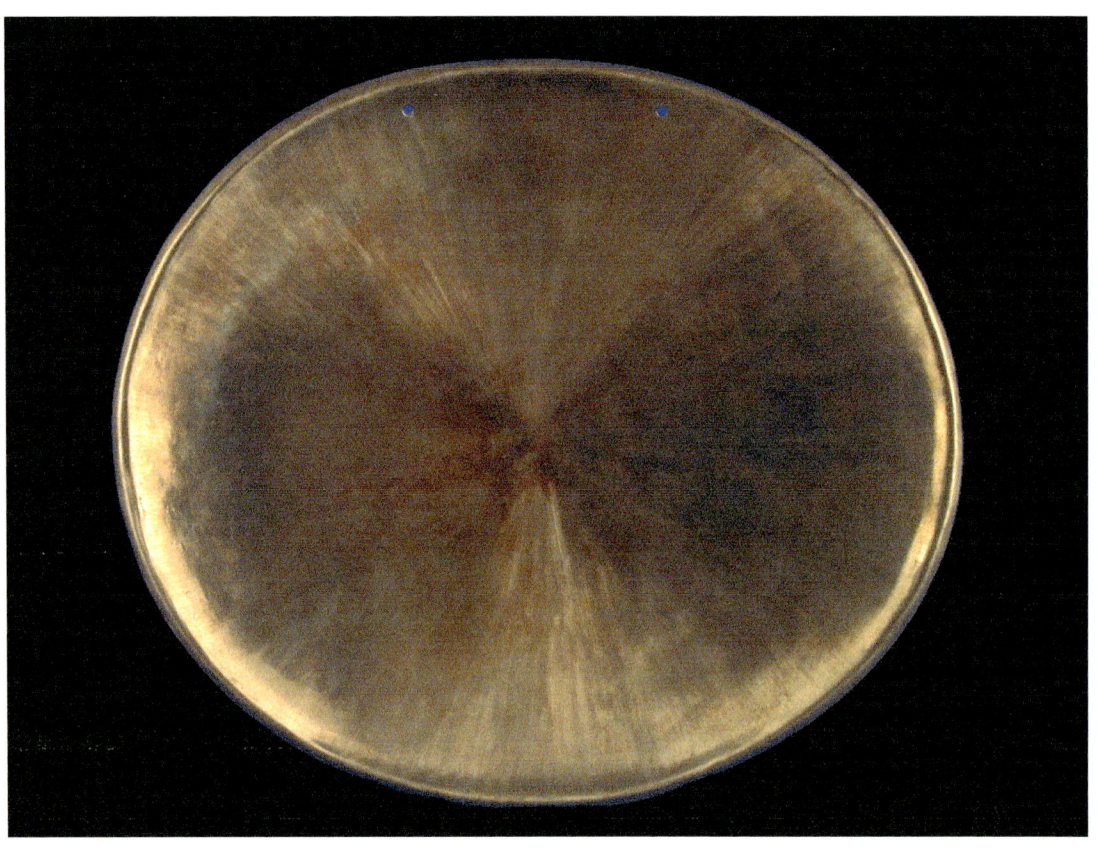

Figure 68 Carchi - Nariño tombac breastplate: 24.5 x 28 x 0.8 cms.

Chapter 13

Isolated finds and problematic Groups

The Great Regional Groups described in the preceding chapters do not comprise the totality of the pre-Hispanic metal objects that have been found in archaeological digs or accidental finds and that form part of the museum collections. In the collection of the Ministry of Culture (Quito) there is a large number of objects without a determined cultural attribution; some of them do not have either provenance information. In the same category there are, however, many objects with exact provenance information; a fact that allows us to take a first look at these isolated finds and problematic groups. In the map we have displayed the provenances of the objects without a determined cultural attribution (green dots) in relation to the provenances of all the other objects (red dots) of the collection of the Ministry of Culture (Quito) (page opposite).

The Coast

The following objects with no determined cultural attribution come from well identified sites in the Ecuadorian coast:

Canton	Types of objects	Frequency	Metal
E-Calderón	Sheet, shank, nose ornament, tweezers	19	Gold, copper
E-Esmeraldas	Axe	2	Gold, copper
E-La Balsita	Nose ornament, sub-labial ornament, shank	14	Copper, gold
E-La Tolita	Figure	2	Gold
E-Palma Real	Shank	10	Gold, silver, copper
E-Rió Verde	Sheet, nose ornament, shank	6	Gold
G-Balzar	Axe	9	Copper
G-Chanduy	Undefined object	3	Copper
G-Milagro	Axe	1	Copper
G-Naranjal	Rattle, tweezers	2	Copper
G-Puna	Nose ornament, sheet, bead	4	Gold, silver
G-San Pablo	Ear ornament	1	Copper
G-San Rafael	Axe	7	Copper
G-Santa Elena	Shank, tweezers	7	Copper
G-Taura	Nose ornament, shank, fishing hook, tweezers, rattle	22	Copper
G-Valdivia	Rattle	1	Copper
M-Bahía	Chisel, rattle, breastplate, tupo, wire, nose ornament	31	Copper, silver, gold
M-Cojimies	Axe	4	Copper
M-El Barro	Bead, plaque, necklace, rattle, tumi, chisel	17	Copper, silver, gold

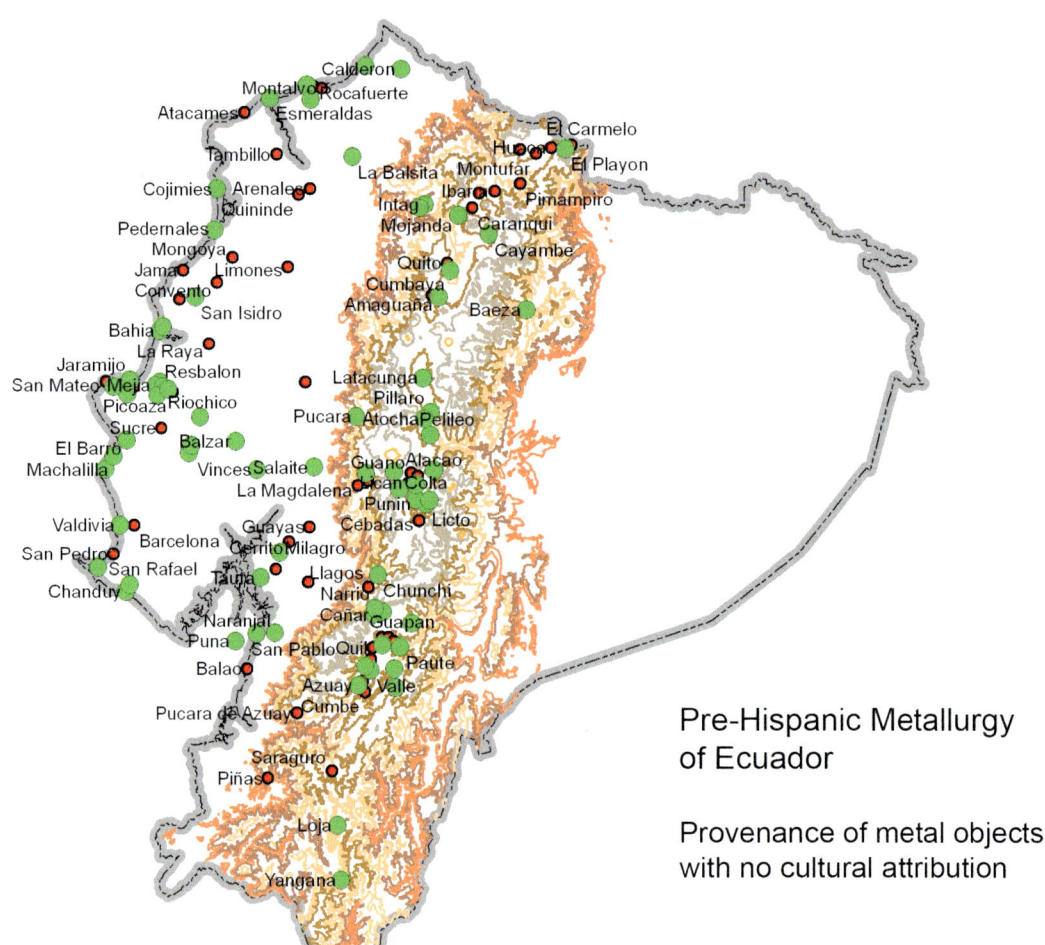

Pre-Hispanic Metallurgy of Ecuador

Provenance of metal objects with no cultural attribution

Canton	Types of objects	Frequency	Metal
M-El Higuerón	Breastplate, bead, nose ornament, sheet	39	Gold
M-Jaramijo	Ear ornament	1	Copper
M-La Norma	Needle	2	Copper
M-La Unión	Chisel	6	Copper
M-Las Chacras	Nose ornament, ear ornament, pendant, needle	20	Gold, copper
M-Machalilla	Necklace	1	Silver
M-Mejia	Bead, applique, tweezers, ear ornament, nose ornament, pendant, rattle	31	Copper, gold
M-Miguelillo	Rattle, nose ornament, sheet, ear ornament	7	Gold, silver, copper
M-Pedernales	Chisel	2	Copper
M-Portoviejo	Undefined object	1	Copper
M-Puerto López	Ear ornament, vase	3	Copper, silver
M-Resbalón	Shank, ear ornament, rattle, sheet	6	Copper
M-Riochico	Chisel, necklace	5	Copper, gold, silver
M-San Isidro	Axe, bowl, nose ornament, chisel, shank	48	Copper, gold
M-San Vicente	Necklace	2	Silver, gold
M-Salaite	Nose ornament, sheet, tupo, breastplate	7	Gold, copper
R-Vinces	Mould, needle, axe-money, chisel, axe, rattle, fishing hook, shank	14	Copper
Total		357	

Especially interesting because of their frequency and the variety of objects and metals used are the concentrations found in six sites of Esmeraldas. Even though some of these pieces are probably part of the La Tolita – Tumaco Early Group, there are also many copper objects that probably fit into the Late Period of the region. Most of the objects from Guayas may be part of the Milagro – Quevedo Group or, less likely, the Manteño – Huancavilca Group. Probably the same can be said of the finds of Vinces in Los Ríos. The objects that come from Manabi are much more varied and, even though they must belong to one of the Regional Groups, it is likely that among them there can be found pieces of the early traditions that started with Chorrera or even before.

A group of 357 unclassified objects in the provinces of the coast is statistically significant and problematic for the definition of a regional panorama of metallurgy. As a matter of fact the number of unclassified objects is larger than some of the collections of the Regional Groups. Even though there are several objects that are very difficult to classify that would require specialised analyses, such as the characterisation of raw materials by isotope or trace analysis, it is also true that a simple formal and typological study would clear many of the doubts.

The Sierra

In the Sierra, from north to south, there are many finds of objects without determined cultural attribution but with precise provenance information, as can be seen in the following table:

Canton	Types of objects	Frequency	Metal
A-Azuay	Axe, nose ornament, breastplate, drill	20	Copper, gold
A-Chordeleg	Sheet, axe, mask, nose ornament, sub-labial ornament, tumi, spear thrower, spear thrower hook	22	Gold, copper
A-Cuenca	Figure	2	Gold
A-Paute	Nose ornament	2	Copper, gold
A-Sigsig	Tweezers, nose ornament, bowl, necklace, shank, bead, wire	29	Gold, copper
A-Valle	Nose ornament, axe	3	Gold, copper
B-Guaranda	Undefined object	2	Gold
B-Pucara	Disc, axe	5	Copper, silver
CH-Chimborazo	Nose ornament, bead, tupo, shank, spear thrower hook, ear ornament, breastplate, spear thrower cover, staff, bowl, tupo, projectile point, axe, rattle	142	Gold, copper, silver
CH-Chunchi	Pendant, nose ornament, breastplate, chisel, tweezers	33	Copper, gold
CH-Colta	Breastplate	2	Copper, gold
CH-Licto	Plaque, bowl, nose ornament	22	Copper
CH-Molobog	Sheet	3	Gold, silver
CH-Penipe	Axe	3	Copper
CH-Punin	Undefined object	1	Silver
CH-Riobamba	Bowl, axe	15	Copper, silver
CÑ-Cañar	Nose ornament, bead, rattle, tupo	19	Gold, copper
CÑ-Cojitambo	Undefined object	2	Gold
CÑ-Ingapirca	Breastplate	4	Gold, copper
CÑ-Narrio	Nose ornament, tupo	5	Copper, gold
CÑ-Pindilig	Nose ornament, pendant, shank, wire	11	Gold, copper
CO-Latacunga	Tupo	1	Copper
CR-El Playón	Undefined object	2	Copper
I-Cerro Junín	Projectile point	10	Copper
I-Intag	Undefined object	1	Silver
I-Mojanda	Head breaker, bolas	3	Copper
L-Loja	Axe	2	Copper
L-Yangana	Axe	2	Copper
NA-Baeza	Nose ornament	3	Silver
P-Amaguaña	Nose ornament, shank, breastplate	7	Gold, silver
P-Cayambe	Nose ornament, shank	3	Copper

Canton	Types of objects	Frequency	Metal
P-Cumbayá	Ear ornament	2	Copper
T-Atocha	Sub-labial ornament, tupo, diadem, nose ornament, breastplate, shank	15	Copper, gold, silver
T-Pelileo	Shank	1	Copper
T-Pillaro	Shank, tupo, nose ornament, breastplate	11	Copper
Total		410	

In the corresponding chapter we pointed out that the collection of the Cañari Regional Group was too small and that it did not reflect the frequent and abundant finds of graves with metal offerings. Nevertheless, it is evident that from the provinces of Cañar and Azuay come many objects with no cultural attribution. In a region that has a tradition of metal working that spans for around three thousand years there must be several periods and styles apart from the well recognised Cañari Group. The study of these objects is very important for the understanding of the early metallurgy of the southern Sierra.

Another interesting and assorted group of objects, made from various materials and containing several different forms, comes from the provinces of Chimborazo and Bolívar. In this region we have only identified the Puruha Regional Group, represented in the collection of the Ministry of Culture by a small sample. As we discussed earlier the metallurgy of this area might be much earlier than supposed and it would be convenient to examine this group of objects with no cultural attribution because other styles, apart from Puruha might be represented in it.

Discussion

Grouping archaeological metal objects into the categories that we have termed Great Regional Groups is a way of recognising the existence of metallurgical traditions whose identity is defined by determined chronologies and spatial distributions, a range of manufacturing and finishing techniques and particular formal, symbolic and iconographic repertoires. When those factors converge coherently we can recognise the existence of Regional Groups, such as La Tolita – Tumaco, Milagro – Quevedo, etc.

In the two previous sections we have seen, however, that both in the Coast and Sierra there are objects whose geographic provenance is known while they do not share a determined cultural attribution. From our point of view the most important task at this stage is to propose interpretative hypothesis with respect to these objects that would otherwise be seen as isolated objects without proper identities. What type of identity could these objects have, since up to now they have only been classified as "undetermined"?

The first hypothesis would propose that these objects form part of unknown or unidentified Phase Components of the Great Regional Groups. Some of the Groups that we have studied have quite long time lives. In these lengthy periods it is probable that considerable stylistic and technological changes would have occurred. Our understanding of a particular metallurgical tradition, the Cañari for example is based upon the appearance of a series of diagnostic traits in well-known groups of objects; if in the same region we find objects that differ from our typological criteria we will be led to believe that they are not Cañari. It may well be that they represent another phase of the Cañari metallurgy, an earlier one for example. It is possible that many undetermined objects found in Cañar and Azuay might correspond to an older Phase Component.

Another hypothesis would propose that there are unidentified Stylistic Components of the Great Regional Groups. At no time have the objects of any given tradition displayed complete uniformity and homogeneity. From one *curacazgo* (chiefdom) to the other there occurred local variations on iconography and form-function of the objects; this variability was further enhanced by the technological differences that necessarily resulted among the diverse workshops and groups of metal smiths. In the province of Chimborazo there are differences between the Puruha objects found in Alacao and those found in Molobog; this happens even though they were buried at approximately the same time. Such differences might help explain some of the groups of undetermined objects; the most prominent example would be found in the provinces of Chimborazo, Bolívar, Tungurahua and Cotopaxi where, apart from the mentioned example, there seem to be many local variations of the Puruha metallurgy. A Tinculpa type breastplate found in the province of Cotopaxi was C14 dated to the 9th century A.D. This seems to indicate that from the beginning of the second millennium copper objects that probably formed part of the Puruha tradition were being made in that area:

Date	Canton	Laboratory Number	Period	Reference
A.D. 1115 ± 95	CO-undefined	Beta 270078	Stylistic Component	Museo del Oro 2010

The third hypothesis proposes that there were other metallurgical groups different and independent of the Great Regional Groups. We would have then Minor Regional Groups that, as the Great Groups, would have a proper identity; that is a distribution area, a chronology and technological, iconographic, formal and symbolic traits of their own. In the present state of knowledge we cannot yet provide such elements for an accurate identification of such Minor Groups. Preliminarily, and mostly to encourage future research, we would like to propose the probable existence of the following Minor Groups.

A Minor Group could have existed in the valley of Quito and the neighbouring areas including probably the high parts of the Cordillera and its slopes towards the

Napo jungle (Baeza). This Group, that does not seem to have had an ample spatial distribution could, nevertheless, have been quite early and an important technological and stylistic repertoire. The finds in the urban area of Quito, especially those of La Florida (Doyon 2002) attest to the use of several alloys and manufacturing techniques and the extended usage of ornaments among the individuals of the elite (Doyon 2002). This Minor Group could have played an important role in the genesis of several styles that thereafter developed in Imbabura, Carchi and Nariño giving rise to the Carchi – Nariño Great Regional Group.

We have a C14 date associated to an object found in Amaguaña (Pichincha) that could belong to the proposed Minor Group. This preliminary information indicates that there was metallurgical activity in the region around Quito since little before A.D. 1000.

Date	Canton	Laboratory Number	Period	Reference
A.D. 850 ± 40	P-Amaguaña	Beta 237169	Minor Group	Museo del Oro 2007

The second Minor Group could have existed in the province of El Oro. On that region, famous for its gold placers the Milagro – Quevedo metallurgy had a late expansion; before that expansion occurred a different style flourished with strong links to the metallurgy of northern Peru (Vicus, Sican). Probably this Group played an important role in the early exchanges between the metallurgies of the Coast and Sierra.

The third probable Minor Group would have flourished in the extreme south Ecuadorian Sierra. In En Loja, and even Zamora, in the eastern slopes of the Cordillera there was a metallurgical tradition derived from the initial nucleus of Putushio (Rehren y Themme 1994). This Minor Group that did not expand beyond a limited area could have been the basis for the development of the Cañari and Puruha metallurgies, thus explaining the close similarities between them.

Undetermined objects can open important research paths. Their correct identification and the correlations that they can help to build might improve our understanding of the processes of expansion and the inter-relations in the pre-Hispanic Ecuadorian metallurgy.

Chapter 14

The Inca metallurgical integration

The Incas arrived into what is now Ecuador by the year 1450; the invasion of the Sierra, commanded by Tupac Yupanqui faces a fierce resistance and only, years after, Huayna Capac completed the conquest of the north (Meyers 1998). The Coast was never directly controlled though it did not escape the economic and cultural influx of the Inca Empire. Inca domination ceased with the arrival of the Spanish and the dismembering of the Tahuantisuyu between 1532 and 1533 (Meyers 1998). We are dealing, therefore, with a very short time span, just over 80 years, that nevertheless marked deeply and definitely the life and culture of Ecuadorian ethnic groups and left a lasting imprint on material culture, including metallurgy. That, together with the fact that probably the influence of Inca metallurgy might have been felt in Ecuador even before the region was invaded. The Incas where an overwhelming cultural force in the north Sierra of Peru since 1400 and it is possible that their metal objects were known in Ecuador before the troops of Tupac Yupanqui made their entry.

An absolute date associated to an Inca object of the collection of the Ministry of Culture confirms the chronology known from historic sources and also validates the dating procedures that we have been carrying out for objects without contextual information:

Date	Canton	Laboratory Number	Period	Reference
A.D. 1560 ± 40	A-Cuenca	Beta 237171	Inca	Museo del Oro 2007

Even though the cultural period preceding the Inca invasion is known in the Ecuadorian archaeology as the Integration Period (A.D. 700 to 1450) the truth is that with respect to the archaeological metallurgy that integration is not at all evident. When we use a term such as integration for a cultural phenomenon that is visible through material culture we can understand from two different points of view: the first sense would indicate that there is an integration of the societies in a regional scale, obliterating the narrow limits of villages and incorporating them into larger political and economic units (Meggers 1966). On the other hand integration can be understood as the configuration of horizons, mainly cultural and perhaps, economic but not political; those horizons would anyhow overflow the limits of the ethnic territories.

However interesting and important it might be, it is clear that this not the place for an in depth discussion of this subject in Ecuadorian archaeology; this is why we shall limit ourselves to metallurgy. From this particular point of view that can undoubtedly be assessed as restricted we can assert that the available evidence shows that:

1. From the moment when the Great Regional Groups acquire their identity, at different times some of them very early, there is an internal integration. This integration is represented, within the territory by the adoption of the use of certain metals and alloys, the implementation of well-defined techniques and the establishment of common forms and iconographies. Such phenomena occurred in some cases in the Late Formative (La Tolita – Tumaco, for example) and in other cases during the Period of Regional Developments (Carchi – Nariño, for example). As a matter of fact, the internal integration of the regional metallurgical traditions took place before the Integration Period.

2. Not even in the years immediately before the Inca invasion there is evidence that there were on-going processes of integration among the regional metallurgical traditions nor, what would be equivalent, among the metal smiths of different ethnic groups. Such integration would have implied a technological, iconographic and functional standardisation that did not exist. There are horizons and trans-regional traditions in the pre-Hispanic metallurgy of Ecuador (spiral ornaments, tinculpas, arsenical bronze, etc.) but those were present before the Integration Period and they did not became stronger or more widespread at that time.

To sum up; metallurgy was already integrated within the Regional Groups before the Integration Period and there never was an integration among the Regional Groups before the advent of the Incas. The real metallurgical integration of pre-Hispanic Ecuador came with the Incas; but even then the integration was never complete, because the coastal Groups escaped from most of the Inca influence (Meyers 1998). However, between 1450 and 1533 the metallurgy of Ecuador attained an unprecedented degree of technological, formal and iconographic uniformity. The Incas brought with them rigid rules concerning the exploitation of mines and placers, the manufacturing processes and, especially, the function and social distribution of metals.

Geographic Distribution

Easily distinguishable Inca objects are found all along the Ecuadorian Sierra. In order to understand the magnitude of the Inca metallurgical industry we would need to take a look at the numerous objects that have been found since the time of the Spanish conquest (Dorsey 1901). We would have to include, as well, the Inca objects in museums and private collections; because the Inca collection of the Ministry of Culture is rather small and most objects have no provenance information. Even so this collection is capable of providing a close idea of the extension and distribution of Inca metallurgy in Ecuador, as can be seen in the following table:

Province	Canton	Number of objects	Percentage of total
Unknown	Unknown	97	53.9
Azuay	Azuay	7	3.9
	Cuenca	23	12.8
	Cumbe	1	0.6
	Huaynacapac	4	2.2
	Valle	1	0.6
Bolívar	La Magdalena	1	0.6
Chimborazo	Chimborazo	4	2.2
	Chunchi	4	2.2
Cañar	Biblian	2	1.1
	Cañar	1	0.6
	Guapan	2	1.1
	Narrio	3	1.7
	Pindilig	12	6.7
	Tambo Viejo	1	0.6
Cotopaxi	Latacunga	2	1.1
	Cotopaxi undefined	1	0.6
Imbabura	Intag	6	3.3
Loja	Loja	1	0.6
	Saraguro	1	0.6
Manabí	Machalilla	1	0.6
Pichincha	Chillogallo	3	1.7
	Quito	2	1.1
	Total	180	100

Setting aside the objects with no provenance information or with deficient information (over 54%) we end up with a completely Sierra pattern of distribution with the sole exception of an object found in Machalilla, Manabi province. This last provenance is not surprising since many Inca objects found their way to coastal regions where they were appreciated due to their technical quality and iconography in spite of the absence of direct political control. Most objects tend to concentrate in the south (Azuay, Cañar and Loja), something predictable because the Inca occupation lasted longer and was more stable in there. However, in the central Sierra (Chimborazo, Bolívar, Cotopaxi) and even in the north (Pichincha, Imbabura), except for the extreme north (Carchi) there are Inca objects in funerary, domestic and ritual contexts.

In the following map we displayed the distribution of Inca objects in Ecuador (green dots) with relation to all the other objects (red dots) in the collection of the Ministry of Culture (Quito):

The Inca metallurgical integration 177

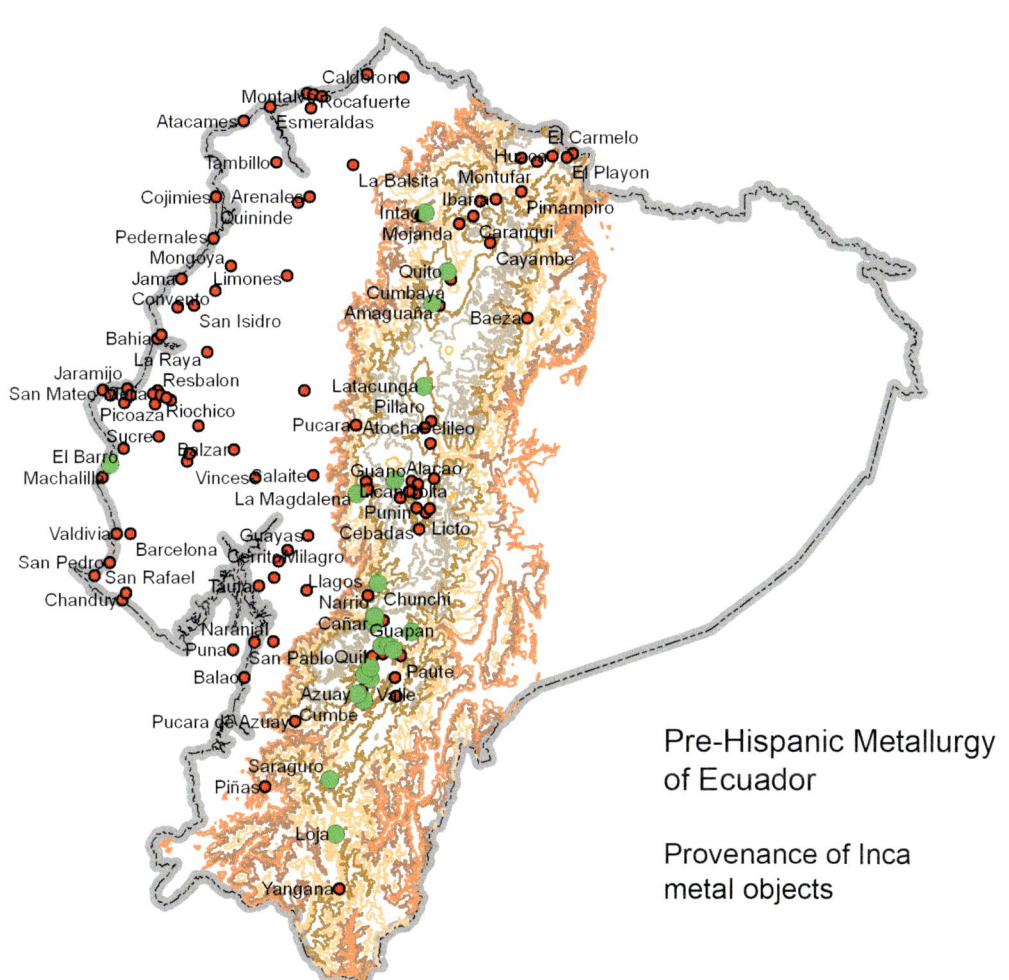

Pre-Hispanic Metallurgy of Ecuador

Provenance of Inca metal objects

Aside from the points discussed in the preceding sections it is also important to note that the Inca occupation did not extend the territorial range of metallurgic production. The Inca objects come from sites where Cañari, Puruha and other objects were being made before, so that the case is the replacement of a tradition by another one. Moreover, in the Sierra the distribution patterns did not change much; where there was no metallurgical production before the Incas (the southern tip of Loja province and the eastern slopes of the Andes, for example) it did not surge during the imperial occupation.

Technology

Inca metallurgical technology is a very complex and broad matter. It coves in South America huge territory and it involves a great variety of metals and alloys, complex manufacturing and finishing techniques and, especially, a scale of production and consumption hitherto unknown in the continent. From the point of view of our study a complete analysis is out of place and it would exceed the research aims. This is why we will only refer to Inca technology in relation to the objects found in Ecuador, mentioning other contexts when it is needed to illustrate a specific case or to explain a trend that emerged elsewhere.

Inca gold is quite scarce in Ecuador; these types of objects seem to have been made in silvery golds with low copper contents that might correspond to natural impurities rather than intentional adding (Barrandon, Valdez y Estévez 2002). Silver is more frequent but there are no composition analyses that might indicate what type of metal was used. Copper and bronze are the most frequent metals; this last one (a copper and tin alloy) apparently was an Inca innovation in Ecuador (Meyers 1998), where it replaced the arsenical bronze, extensively used in previous times. Inca bronzes have up to 6.8% Sn, a proportion that does not correspond to the natural impurities of the copper minerals (Verneau et Rivet 1912). Copper and bronze objects were frequently gilded, probably by sheet gilding. There are no evidences of the use of tombac or depletion gilding.

Hammering and casting were systematically employed on determined types of artefacts, thus revealing the strict application of manufacturing rules imposed by the State. Usually hammering started with cast pre-forms, thus making production more efficient and quick. Hammered tools and weapons were hardened by forge; hot objects were constantly hammered during cooling down obtaining the desired degree of deformation and a hard structure (Meyers 1998). To make containers the sheets were hammered to the desired thickness and then they were assembled by welding together the edges (Meyers 1998). Welding was used also to make tupos and copper, gold and silver objects from small components. Gold and silver sheets were used also for plating, sometimes gold plating on silver or gold and silver on wood, as with staffs. Hammered objects were usually decorated by embossing, working on both surfaces and engraving (also used for cast objects). An innovative technique for decoration was inlaying metal on metal.

There are no evidences of lost wax casting, probably because it is a slow and laborious technique not apt for mass production. Instead, sophisticated moulds were used; there were bivalve moulds that are very efficient but have restrictions with respect to the geometry of the cast object and composite moulds made with several parts that fit together and allow for the extraction of the cast objects without getting damaged. The possibility of re-utilisation of composite moulds made them ideal for the production hundreds of votive figures (conopas). Sometimes these figures were made in separate parts (trunk, extremities, head-dress) that fit together; this manufacturing technique made massive production even more efficient.

Even though the Incas imposed their technology and stylistic patterns to the ethnic groups of Ecuador, displacing ancient developed traditions, it is also true that they also adopted some of the techniques and iconography giving rise to half-breed styles, as it happened in Cañar, Chimborazo and Loja where a mestizo cultural phase with a strong Inca influence is evident (Fresco 1984 and Collier y Murra 1982)

Typology and classification

Inca metallurgy comprises a wide range of objects, adornments, weapons and tools that throughout the empire acquired regional characteristics creating different local styles. This is not the place to deal with such a variety; as with technology we will only consider the form and function categories that appear in the Inca metallurgy of Ecuador and specifically those represented in the objects of the collection of the Ministry of Culture (Quito), summarised in the following table:

Function	Form - Representation	Frequency	Percentage
Pin	Simple - total	6	3.49
Shank	Simple – total	14	8.14
Arivaloid vessel	Simple – total	1	0.58
Bolas	Simple – total	5	2.91
Bracelet	Simple – total	2	1.16
Rattle	Simple - total	10	5.81
Chisel	With strangled heel	2	
	With straight heel	2	
	Total chisels	4	2.33

Function	Form - Representation	Frequency	Percentage
Pendant	Simple	3	
	Head shaped anthropomorphic	2	
	Circular	1	
	Bird shaped	1	
	With dangler	3	
	Total pendants	10	5.81
Bead	Cylindrical - total	1	0.58
Figure	Anthropomorphic feminine	6	
	Anthropomorphic masculine	5	
	Zoomorphic	3	
	Anthropomorphic	2	
	Shaped as a feline	1	
	Masculine miniature	2	
	Total figures	19	11.05
Axe	Simple	5	
	With strangled heel	5	
	Shaped as an anchor	1	
	With openwork heel	5	
	With drilled heel	2	
	Total axes	18	10.47
Axe anvil	Simple - total	1	0.58
Sheet	Rectangular - total	2	1.16
Ingot	Simple - total	2	1.16
Nose ornament	Simple	3	
	Circular	1	
	Total nose ornaments	4	2.33
Ear pendant	Annular - total	2	1.16
Breastplate	Trapezoidal - total	1	0.58
Projectile point	Simple total	1	0.58
Head-breaker	Simple	2	
	Star shaped	7	
	Total head-breakers	9	5.23
Drill	With zoomorphic figure - total	1	0.58
Tumi	Simple	2	
	With zoomorphic figure	3	
	Zoomorphic	2	
	With strangled heel	1	
	Total tumis	8	4.65
Tupo	Simple	46	
	Miniature	3	
	Total tupos	49	28.49
Vase (kero)	Simple - total	2	1.16
Total		172	100.00

The features displayed by this form – function distribution are to be expected in a tradition that was subject to rigid social rules. The most frequent adornment are simple tupos, as those that could be used by common women, pendants and rattles (40.11%), while other ornaments, presumably restricted to the elite (bracelets, nose ornaments, ear pendants, breastplates) are quite uncommon (5.23%). Equally scarce are special and elaborate utensils, as the arivaloid vessel and vases (keros) that account only for 1.74% of the sample.

Tools and weapons are very important since they were probably used by many groups of the society; pins, bolas, head-breakers, axes, axe-anvils, chisels, projectile points, drills, tumis and spear-thrower hooks (n. d.) represent 30.82% of the sample. The other abundant type of object is the votive figure, probably extensively used by everybody in compliance with the precepts of the State religion (11.05%).

In general terms it is a Group massively oriented towards simple adornment of common people and the efficient production of tools and weapons. Secondarily the industry produced sumptuary items as those used by the elite and the State officials.

Most Inca metallurgy was in use, not buried in graves, at the time of the Spanish conquest and thus it was unavoidable that the vast majority got lost and was melted in the furnaces of the Conquistadors, especially the gold and silver objects. Inca elite graves in Ecuador, as the one found by Dorsey (1901) in Isla de La Plata, are rather exceptional. Meyers (1992) mentions other objects not represented in the collection of the Ministry of Culture like picks, hoes, hammers, wedges, gouges, pots, mirrors, tweezers, forks, bells, tassels, plates, figures of fruits, pitchers with legs, elongated plaques and axe-maces.

Figure 71 Inca silver ceremonial vase (*kero*): 17.8 x 16.7 cms.

Figure 72 Inca silver arivaloid bottle: 39 x 29 cms.

Figure 73 Inca gold and silver votive figures: 2.9 x 1 x 1 and 2.9 x 1 x 1 cms.

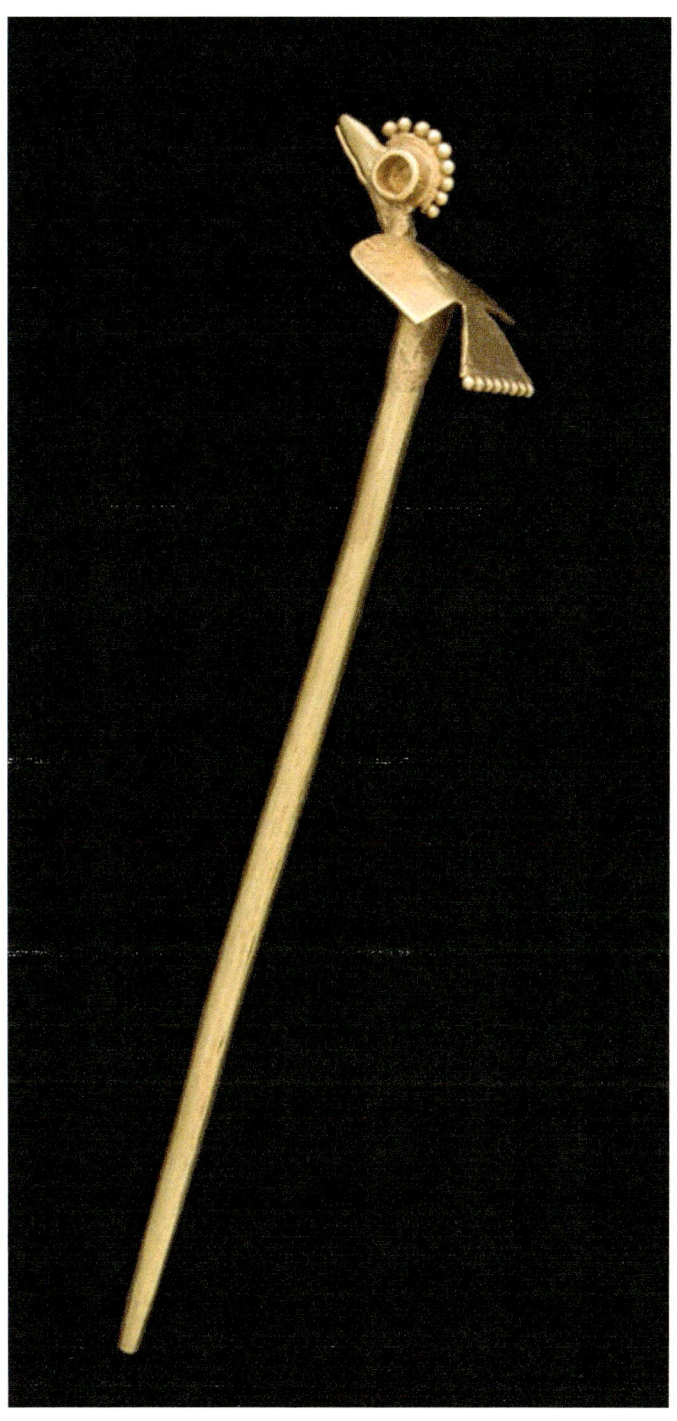

FIGURE 74 INCA GOLD *TUPO*: 13.5 x 1.6 x 3 CMS.

Figure 75 Inca bronze axe: 8.4 x 8.2 x 0.3 cms.

The Inca metallurgical integration

Figure 76 Inca bronze head breaker: 2.6 x 9.4 x 1.5 cms.

Chapter 15

Iconography and symbolism in metallurgy

It is important to explain why the study of symbolism and iconography is carried out jointly for all the Great Regional Groups and not separately for each, as was done for the spatial, chronological, technological and functional aspects. The first, and most important, reason is that symbology and iconography, in spite of the fact that they have a material base that is expressed in spatial and temporal concrete referents, have the ability to overcome time and space frontiers. As they extend beyond the geographic and chronological limits within which they emerged, they tend to create relations of forms and meanings; this is, more than any other feature, the one that links together the Great Regional Groups. The existence of these relations and networks supports the presence of common motifs, signs, symbols and icons. We propose to show those features, common to all Groups, and discuss their concrete and particular expressions in each one.

The second reason is that some Groups are represented by small and limited samples, so it is difficult to say something significant. And the last reason, which is directly derived from the first one, is that the great quantity of common themes and treatments of the symbol and the icon among the diverse Groups would have forced us to a large dose of redundancy. Much of what we would have said in the La Tolita –Tumaco chapter would have been repeated in the Jama – Coaque chapter and so on.

Once this has been explained all we have to do is to describe how the following discussion is organised. First we will deal with the general form of the objects and for that purpose we will start with geometric shapes such as circles, rectangles, cylinders, etc. Next we will approach representations (anthropomorphism, zoomorphism, etc.); then the theme of symbolism with an emphasis on dualism. At the end we will structure a global discussion taking into account all the subjects. In every case we will handle general and particular aspects illustrated with examples. Finally we need to warn that the statistics cited might be affected by deficiencies in the classificatory processes of the collections; for example, not all the objects with a certain shape have been described in the same way, sometimes the description of a common form was omitted because it was considered too obvious.

Among bi-dimensional geometric forms the circle and its related shapes (oval, ellipse, annular, semi-circular, semi-elliptical, arc) are the most frequent. Approximately 15.4% of the objects from all groups have that general form; if we add the tri-dimensional forms derived from the circle (sphere, globular, semi-globular, semi-spherical, conical, cylinder, truncated cone, bi-conical, tubular, spool, half spool) we have an additional

7.6%. Up to 23%, almost one quarter of the objects of the collection have circular forms or forms derived from the circle in two or three dimensions.

The geometry of the circle and its derived shapes are equally frequent in metallurgical styles of other regions of the continent, but the numeric data is almost never mentioned so it is impossible to properly esteem that predominance. Moreover, the predominance of the circle and its derived shapes is not limited to the general form of the objects; the circle is present as a decorative element (openwork, repousse, granulation welding) on circular pieces (circles on circles) and other geometrical forms (circles on rectangles, triangles, etc.). The circle also makes part of the configuration of objects that have other general forms; rounded corners, endings, ornaments and danglers are elements that introduce the circular shape in objects that are not circular.

Circular or close to the circle, in two or three dimensions, are most breastplates, including the tinculpas, many ear pendants and ear ornaments, some nose ornaments, pendants, beads, endings of tupos, spear throwers and staffs, helmets, bowls, crowns, diadems, many sheets and all the rattles. If we were to include those objects, which have not been classified as circular, that geometrical shape may well approach 50% of the sample.

Another frequent shape in Ecuadorian metallurgy, the spiral is undoubtedly a derivation of the circle, as well as its projection in three dimensions, the spring (a concentric circle or a circle that unfolds). Bray (1971) recognises in the spiral one of the distinctive marks of the metallurgical tradition of this country. Our statistics indicate that the spiral is present only in approximately 2% of the objects of the collection. As with the circle and all the other geometric shapes, these percentages are lower than expected; nevertheless the spiral, however notorious it might be, is a frequent motif only in certain objects such as the Milagro – Quevedo ear pendants. As a decorative element the spiral is more popular and exists in all Groups.

Among curvilinear shapes, not proper circles, the semi-lunar shape is quite important, though according to statistics that shape would be scarce. A related shape is the anchor; both are especially frequent in Carchi – Nariño, but they appear also in the other Groups and they are, no doubt, the shapes of the blades of axes and tumis.

Rectangles and related shapes (squares, bands, elongations, trapeziums, rhombus, prisms) are very few; less than 2.5% of the sample. There are, nevertheless, certain large and elaborate rectangular objects, like breastplates, sheets and sets of textile appliques. The other geometric shapes (triangles, pyramids, crosses, stars, lenticular) are almost insignificant in the sample.

From a geometrical point of view the pre-Hispanic metallurgy of Ecuador can be seen as an exploration of the circle; circles that are divided, stretched, projected, multiplied and unfolded in two and three dimensions. There is very little left for the opposite shape, the rectangle that was not handled with the same interest or enthusiasm.

Anthropomorphism is present in all Groups; an anthropomorphic cosmos demands the overwhelming representation of the human body. Anthropomorphic representations can involve complete bodies, as in La Tolita – Tumaco and Inca bodies, just the head or other parts of the body like arms, legs, ears, noses (also in La Tolita – Tumaco). In these categories of anthropomorphism we found just over 2% of the sample, but this percentage can be greater; just one type of object, tinculpas, is mainly a development of anthropomorphism and there are ear pendants, nose ornaments, breastplates, spear thrower hooks, staffs, etc. decorated with anthropomorphic figures.

Anthropomorphic representations are, generally, frontal, hieratic and devoid of movement. Occasionally coca chewing is represented (Manteño - Huancavilca, Carchi - Nariño and Puruha). Side representations are uncommon, they are found especially in Cañari ear pendants that depict a winged person with a tail walking. Movement of the figures is suggested in Cañari in other ways; some gold ear pendants with silver backings depict a central head around which other human heads, birds and feet seem to spin. Most tri-dimensional figures (La Tolita – Tumaco, Puruha and Inca) maintain the hieratism of flat figures; there are few exceptions, like the Manteño –Huancavilca figure crudely cast in copper representing a person with almost no arms seated with its legs crossed. The most obvious exception is the Bahia raft, a votive object that represents a navigation scene; two persons row, one guides the raft and the central person remains static. In spite of this last attitude the raft unequivocally communicates a sense of movement.

Genre is not generally explicit except for Inca votive figures that clearly differentiate sexes. Some aggressive gestures in the faces of Jama - Coaque, La Tolita – Tumaco and Carchi – Nariño might be interpreted as masculine, but there is no clear treatment of genre i those representations. A tubular bead with a feminine figure in Puruha and a masculine figure with exaggerated sex organs in Manteño – Huancavilca are among the few exceptions to this rule; it must be noted that this last figure could be the product of Inca influence.

Anthropozoomorphism changes from Group to Group. While the combination of features of animals and humans is extremely frequent in La Tolita – Tumaco and has some importance also in Jama – Coaque, it is virtually inexistent in the other Groups. The remarkable exceptions are the breastplates, pendants and ear pendants of the tinculpa type. Tinculpas exist, in high or low quantities, in all metallurgical Groups (Jijon y Caamaño 1920). They are extremely frequent in Carchi – Nariño and Manteño - Huancavilca, very scarce in Cañari and Puruha and appear in different quantities in the other Groups, but they are always present. Invariably from the centre of the circular shape there emerges a head (sometimes two or three) and that head, in over half of the cases is a human-animal. Tinculpas, due their diffusion and the consistency of their representation pattern, are the most important iconographic tradition of the pre-Hispanic metallurgy of Ecuador and one of the most interesting research subjects. Undoubtedly it is an icon whose meaning crossed ethnic and cultural borders and was recognised in many different social groups for a long time.

Zoomorphism is another of the recurrent depictions in Ecuadorian metallurgy. In our sample 3.6% of the objects have a general zoomorphic depiction, animal features or are decorated with animal figures. Even though this is a low percentage for a depiction that is declared as recurrent, we must not forget that there are many zoomorphic objects that were not classified as such. Among the animals depicted the first place is for birds (without specifying species) and in descending order of frequency monkeys, snails, felines, owls, insects, crustaceans, reptiles and deer. Parts of animals, such as fangs and claws are also depicted. There can be other animals that we have not recognised, partly because some are represented in a very schematic way. Among the objects examined we were able to pick up the features of very schematic snakes in La Tolita – Tumaco objects; in the *Museo del Oro* collection there are several fishes and a frog.

Undoubtedly the most important type of animal is the bird. In certain Groups birds are quite remarkable and they play key roles; in Carchi – Nariño the birds that are depicted on pendants and breastplates are probably condors, in Manteño – Huancavilca plaques and breastplates exhibit the figure of what is possibly an eagle or a hawk with spread wings. In both cases those are emblematic and prominent figures, even more than anthropomorphic representations.

Monkeys are quite important; they are not present only in Carchi – Nariño, as could be presumed, though they are more frequent there. We found them also in Manteño – Huancavilca plaques, Puruha spear thrower hooks and La Tolita – Tumaco pieces. Felines, probably tropical jungle jaguars, are prominent in La Tolita – Tumaco and Jama - Coaque where their depiction is centred on their jaws with great fangs. Other animals have very limited frequency and importance.

The last depiction that we shall deal with is phytomorfism. Around 2.3% of the simple is phytomorphic or has features or ornaments of this type. Complete plants are rare but flowers with or without stem, leaves and fruits, especially gourds are depicted. Most depictions of his kind come from La Tolita – Tumaco, but there are some in Puruha as well.

All depictions together (anthropomorphic, anthropozoomorphic, zoomorphic and phytomorphic), leaving aside possible omissions, just amount to 10% of the collection. Probably a detailed research would double this percentage, but it is still relatively low.

What this suggests is that metallurgy, all Groups included, is not especially rich in iconographic depictions. There is a general trend towards simple shapes, preferably circular, with little attention to realistic representations of the type that we can recognise.

There are, of course, several objects that fall outside of this trend and exhibit a rich and complex iconography; this is the case, for example, of the "Golden Suns". These large, emblematic pieces must have had special roles that required such degree of elaboration; while for most objects, especially the small one that does not seem to have been the case. In this order of ideas it must be noted that there is some variation; La Tolita – Tumaco is the richest Group in iconographic terms while Cañari and Puruha are quite plain.

Dualism, that is the conception of society and nature as an orderly and equilibrated set of opposite principles, is a salient feature of Andean thought (Lleras y Gómez 1999). The expression of dualism in pre-Hispanic metallurgy was clearly described by Gómez (2003) for the Carchi – Nariño Group. This author showed how the interplay of concave and convex, full and empty, bi-colour and bi-texture, bi-metallic and double headed tinculpas communicate dual conceptions that were characteristic of the social organisation of the groups that lived in the plateaux of Carchi and Nariño and that are still present among their descendants, the present day Pasto Indians.

Dual iconography is present in the other metallurgical Groups, besides Carchi – Nariño. In each Group its concrete expression is different and specific. The use of gold and platinum in La Tolita – Tumaco allowed the manufacture of bi-colour pieces and figures; occasionally, as with some nose ornaments and ear pendants, each half of the object has a different colour. This dual division is also represented using the front and the back of the objects or creating intricate designs for each colour, as with masks. Another form of expressing dualism is through rattles with two faces and ear pendants with diverging spirals.

In Jama – Coaque dualism is less evident, except for diverging spirals, though there is certain interplay between the colours of gold and tombac. The introduction of silver in Bahia shapes dual expressions based on the colours of this metal and gold; pairs of identical objects, one in silver, one in gold, were probably part of a unique set of grave offerings.

In Milagro – Quevedo double spirals are very frequent, but the expression of dualism is not limited to that. The use of gold, copper, silver and gilded copper permitted the interplay of different colours that might have been arranged in binary sets of opposites within the grave offerings, as suggested by the contents of the grave of the "Guayas Chief" (Meggers 1966).

In Manteño – Huancavilca the use of silver, gold and copper, was even more popular and served as the basis for binary colour oppositions. It is in this Group that the most explicit iconography is found; in breastplates, pendants, plaques and axes it is usual to find pairs of human figures, and less frequently pairs of monkeys or bulges (those last also in quartets). This couple of identical "twins" is sometimes holding hands or supported on a bar that divides in four a scenic space. There are also the well-known double spirals, double and triple tinculpas (as in Carchi – Nariño and Puruha); in several studies of Andean cosmology (cited in Lleras y Gómez 1999) it has been shown that the triad is part of the dual system since it allows for the construction of systems of double oppositions.

Dualism is present in as the interplay of surface colours in Puruha and Cañari; in these cases identical objects of gold and silver were deposited in pairs in the graves of elite persons (Saville 1924, Uhle 1922). But they also sought by means of sophisticated techniques to integrate both colours in the same object; rectangular gold pendants with silver corners, golden semi-lunar nose ornaments with silvery centres, spear thrower gold covers with silver and occasionally copper ribbons (thus introducing the triad in the composition).

The dualism of Inca metallurgy is not a new discovery, it has been referred by those who have studied it and it was known since the time when the Spaniards described the looting of Cuzco. In the sample of the Ecuadorian collection the most remarkable feature is the use of gold and silver for the massive production of votive figures that were usually deposited in pairs; there is also a copper figure of an individual with two opposite faces. Meyers (1998) mentions dinnerwares of gold and silver and depictions of life-size fruits and plants in gold and silver.

Inca cosmology that linked gold with the sun and silver with the moon has been extended to many other Native American groups without enough evidence to prove if that scheme really dominated their cosmology or without verifying if that idea really existed outside of the Inca thought. The fact that dualism was the logical basis for thought almost everywhere in pre-Hispanic America does not imply that the sun-moon dualism was equally extended; dualism is based on many other oppositions, mainly that between masculine and feminine.

This argument, of the sun and moon, is precisely the one used by Ledergerber (2004) in one of the few iconographic studies published on Ecuadorian metallurgy. According to the author the geographic factors of Ecuador converge with cultural diversity to configure a relation and influence of the sun and the moon: *"...twelve hours of light/gold, represented by jaguar/man;"* and *"...other twelve hours of darkness/silver, represented by caiman/snakes,"* (Ledergerber 2004). In spite of the literary appeal that the text may have, it is rather difficult to accept that such general interpretation may be constructed on the basis of a study of a limited and de-contextualised sample.

It is possible that other iconographic, symbolic and mythological themes are present in the vast universe of Ecuadorian pre-Hispanic metallurgy; concepts like transformation, sexuality and fertility, shamanic flight might be depicted in forms and objects that are not yet well studied or understood. However, as long as there is no firmer basis to draw interpretations it is better not to propose adventurous hypothesis.

Chapter 16

Synthesis

By the time of the arrival of the Spaniards in Ecuador, metallurgical traditions had been developing for around three thousand years in this territory. What was a flourishing industry, successful in handling several metals and alloys and masterly in the manufacture of thousands of adornments and utensils, was brutally attacked. Looting and stealing of gold and silver went on unceasingly until there was nothing on the surface to take. Then came the hurried search for graves and only when this source was exhausted the exploitation of mines and placers was organised.

It was necessary to wait until the 19th century when some foreign travellers and Ecuadorian scholars began to take interest in the archaeological metal objects that by then were still being looted in the graves of La Tolita Island, La Tolita, Carchi, Azuay and Cañar. Slowly a conscience about the cultural value of these objects emerged, while some pioneer scientists began analysing the objects and started to establish relations among them, the pottery pieces, the architectural vestiges and the material equipment of other cultures in the continent.

Since the end of the 1930s there are structured metallurgical and archaeological studies, as those by Paul Bergsoe whose conclusions are still valid nowadays. By the 1940s the Banco Central del Ecuador started its collection of metallic objects that grew steadily, in part through the acquisition of old private collections, thus preserving, researching and divulgating over five thousand metallic archaeological objects.

The construction of the knowledge about the metallurgy of this territory was a slow process that was nourished by archaeological digs and the study of museum collections; both activities were also carried out in southern Colombia thus contributing to the metallurgy of Carchi and La Tolita. This is an on-going process that has built an important corpus of data that, with the exception of the museum layouts, has not been compiled in a global synthesis. Even today no agreement has been reached among the archaeologists working in this field with respect to the basic aspects (space, time, technology, etc.) that would allow the formulation of a paradigm in pre-Hispanic metallurgy. On the contrary, there is an overwhelming emphasis on the factor of foreign influences that ends up concluding that the metallurgical development of Ecuador was the result of diffusion, either from Peru or Colombia.

Ecuador occupies a territory on the Andean Cordillera where there is an intense magmatic activity; in these areas there are important deposits of copper, gold, platinum, silver and arsenic, the main metals used by pre-Hispanic Indians. Even though those deposits are not especially rich they are quite extensive and they were enough to supply the demand of metals at the time. The characterisation of raw materials has allowed us to conclude that in the coast they used placer gold and in the Sierra mine gold. We still need to work out the topic of the provenance of the copper used in the coast, because in this area there are no deposits of this metal. Platinum is a topic of its own; its presence in the placer golds of the north coast definitely marked the metallurgy of that area.

Even though mining activity, both in mines and placers, must have been intense we still have no concrete confirmed information about mining techniques, infra-structure and processes. We know though, that many early colonial exploitations, like the Santa Bárbara mines, were the continuation of pre-Hispanic indigenous works.

In the continental context Ecuador has many evidences of early metallurgy. The available information indicates that there were, at least, three very early foci (1500 to 1000 B.C.) where metalwork developed; those are the province of Loja in the south Sierra (site of Putushio); the Santa Elena Peninsula and Manabi in the central coast (sites of Los Cerritos and Salango) and the north coast (site of Las Balsas). In two of these sites (Putushio and Las Balsas) we can establish complex sequences of dates and finds that configure regional metallurgical traditions.

The available information suggests that in the Middle Formative there was already an important metallurgical production whose tracks have not yet been identified in the museum collections. This theme deserves an exhaustive revision of many objects that might have been incorrectly classified and also a field project aimed at solving the riddle by means of systematic digs.

The La Tolita – Tumaco Regional Group developed in the north coast, province of Esmeraldas and north Manabi, and in the south Pacific coast of Colombia up to the bay of Buenaventura. The initial date for the Early Period is around 500 B.C. and we have evidence that it lasted until around A.D. 200. A second stage, the Late Period, appears from A.D. 700 and continues until the conquest in A.D. 1500. La Tolita – Tumaco technology is probably the most varied and complex in this territory; the most remarkable feature is platinum handling by means of several variations of sintering, thus allowing smiths to create solid objects, sheets, inlays and platings with this metal. Hammering of gold, tombac and copper, welding, gilding, lost wax casting and precious and semi-precious stones inlaying were also practiced. The range of forms of the La Tolita – Tumaco metallurgy is dominated by a great variety of small objects and miniatures especially face nails; there is a great emphasis on the adornment of the head and face and less attention to the trunk and extremities. Utensils and tools are also important.

The Jama – Coaque Regional Group is concentrated on a few sites in the centre of Manabi, central coast. There are few evidences to precisely define its chronology that we might estimate between 400 B.C. to A.D. 600. I technological terms this Group is characterised by the predominance of hammering on silvery golds, embossing and plating of organic materials with god sheets. A large proportion of the objects are nose ornament and ear pendants, as a rule objects are quite small.

The Bahía Regional Group has a very wide distribution covering almost all the north and central coast especially, though not exclusively, by the seaside. The very scarce archaeological evidence suggests a chronology between 100 B.C. and A.D. 1500. This Group introduced silver into the metallurgy of the coast; hammering, mechanical assemblage and welding were used for the production of complex objects. Gilding of copper surfaces, embossing and semi-precious stones inlaying were also practiced. Bahia objects are mainly small ornaments for the face and head and, in lesser quantities, large and complex pieces like silver chest guards and votive figures.

Even though it is usually considered that the Guayas – Daule basin was the traditional territory of the Milagro – Quevedo culture, this type of metallurgy is distributed along a larger territory that includes parts of the north and central coast and some isolated sites in the central Sierra. The chronology of Milagro – Quevedo covers the period from A.D. 200 to A.D. 1600. In this time span there flourished a copper and arsenical copper metallurgy whose production volumes were unequalled at the time; large staffs, giant axes and thousands of axe-monies poured out of moulds in workshops that also made delicate gold, tombac, silver, gilded and silvered copper objects. The domestic production of utilitarian copper objects was important in the Milagro society; specialists were sometimes buried with the tools of their trade. In terms of form and function the Milagro-Quevedo metallurgy is divided into two great groups; adornments, mainly nose ornaments and tools and axe-monies that characterise this Group.

The Manteño – Huancavilca Group is disseminated mainly on the central coast, provinces of Manabi and Guayas, but it is usual to find Manteño pieces in the Guayas – Daule basin. Its chronology, deducted from a considerable number of C14 dates, covers the span from A.D. 700 to A.D. 1600. The technology shares many traits with the Milagro – Quevedo Group, but there is a more intense use of silver that was hammered and cut with precision and copper wire, an element that was shaped and assembled to make complex pieces. There are thousands of axe-monies and other copper objects; a far reaching tradition grew and managed to extend its techniques and formal repertoire to the west coast of Mexico. In Manteño metallurgy the face and head ornaments are not so important, most objects are utensils and tools. Another important feature of this Group is the existence of central Andean forms like tupos and tumis.

The Puruha Regional Group is distributed on the central Sierra, provinces of Chimborazo, Bolívar and Tungurahua. Its chronology is not completely defined but, on the basis of C14 dates, it is estimated between A.D. 200 and 1500. Puruha metal smiths made an extensive use of gold, silver and copper, mainly by hammering and cutting, techniques that they mastered. The most remarkable technological feature is the combination of gold and silver, and copper to a lesser extent, to make bi-metallic objects by means of a technique that combined hammering and welding. Puruha objects include large and flashy body ornaments, giant tupos as well as ritual objects and tools.

Cañari metallurgy, of the Sierra provinces of Azuay and Cañar, is well-known from the time of the conquest and has been extensively looted. The archaeological digs have obtained some C14 dates that, together with relative dating, have allowed locating this Group between A.D. 900 and A.D. Cañari technology shares many features with Puruha, among them the manufacture of bi-metallic gold – silver piece and the plating of spear throwers and staffs with sheets of gold, copper and silver fixed with nails. Arsenical copper was important to make tupos and ornaments made by casting and finished by hammering. Cañari objects include, besides large adornments like ear pendants and circular breastplates of gold, silver and gilded copper a large quantity of copper tools and weapons like axes, axe-picks, head breakers, whistling projectile points and chisels.

In the plateaux of Carchi and Nariño, north of the Chota canyon and south of the Patía valley in Colombia, there are many finds of Carchi – Nariño objects. Probably the early phases of this Group are to be found in the valley of Quito where the La Florida site has provided abundant material and information. From there it may have disseminated to the north through Imbabura and the Chota valley, where in sites as Pimampiro there are unusual concentrations of finds. In this region we have identified: the Early Period with the Yacuanquer and La Cruz styles (A.D. 100 to A.D. 600); the Middle Period with the Southwest Nariño – Carchi style, Capuli and Piartal sub-styles (A.D. 800 to A.D. 1500) and the Late Period with the North Central Late style (A.D. 1500 to A.D. 1700). Technology varies from style to style, but in general terms it is characterised by the use of tombac, fusion and depletion gilding of hammered pieces. A special technique, shared only with Puruha and Cañari is zoned scraping that allowed producing bi-colour and bi-texture surfaces. There are few tools, most objects are body ornaments, especially for the head and face; in the Intermediate Period most ornaments seem to have been made for the exclusive use of elite persons while in the Late Period objects seem to have been worn by common people. A very special type of ritual object is the spinning disc.

A large number of metallic objects have not been classified within the Great Regional Groups; they are varied artefacts that appear in large quantities in many sites of the Sierra and coast. We have identified preliminarily some concentration foci of this type of finds in the provinces of Azuay and Cañar, Chimborazo, Bolívar and Tungurahua and the valley of Quito and neighbouring areas. We propose that these objects might represent categories like Phase Components or Style Components of the Great Regional Groups or independent Minor Regional Groups.

With the Incas there is a true metallurgical integration in the Ecuadorian Sierra that never before existed. The arrival of the Incas, around 1450, marks the introduction of a new alloy, tin bronze, and a massive and organised production that standardized materials, tools, forms and decorative motifs. The coast did not escape from this overwhelming influence. Lost wax casting was replaced by casting in bi-valve and complex moulds and cold working with annealing was replaced by forges. Inca objects in Ecuador are mainly a few simple ornaments like tupos and rattles and abundant tools and weapons like axes, chisels, hoes, gouges, projectile points, head breakers and drills. Large gold and silver objects for elite use are exceptional.

All the Regional Groups share common aspects with regard to iconography and symbology and, likewise, each has its own particular features. With relation to form the most remarkable feature is the preponderance of the circle, a geometric shape that was intensely explored to obtain many different compositions. The spiral is a characteristic and disseminated variant of the circle. Rectangles and the other geometric shapes are less important. The world of depictions is dominated by anthropomorphism; human figures are usually hieratic, frontal and devoid of movement; there are rare exceptions to this rule. Anthrozoomorphism is rare, except in La Tolita – Tumaco and the numerous objects of the tinculpa style. Among zoomorphic depictions birds are predominant, followed by monkeys. The other significant set of depictions is phytomorphic; flowers, stems and leaves appear with certain frequency.

The most evident iconographic topic is dualism; it is represented in all Groups, though with different intensity and in diverse ways. Dualism is expressed through different metals that give identical objects, or parts of an object, contrasting colours. Another way of expressing dualism is through double spirals, multiple faces in tinculpas and the interplay of concave and convex, full and empty. Manteño – Huancavilca iconography is the most expressive in this sense; the icon of the "twins" that portrays two persons side by side holding hands is frequent in diverse pieces. Probably other iconographic and symbolic themes are present in the wide universe of forms and representations, but finding them is a hard task that should avoid rushed speculation.

Chapter 17

An interpretative proposal for the development of metallurgy in Ecuador

Any global interpretation of the pre-Hispanic metallurgy of Ecuador that makes use of the available information while providing ideas to guide research must, in the first place, free itself from certain adverse prejudices. Almost with no reason there has been an insistence on the role of "transition area" of Ecuador in the history of the development and expansion of South American metallurgy. It is almost impossible to avoid unconvincing explanations about how the influence of Wari, Mochica, Vicus or Sican entered the coast or Sierra of Ecuador or how, later on. This influence came from the north, from Calima, Quimbaya or San Agustin. It is true that certain degree of diffusion and exchange of technical knowledge or iconography occurred throughout America from the Early Peopling and that metallurgy did not escape this phenomenon, but it is also true that this image of a purely receptive area is not coherent with what we know at present.

In order to construct a different argument it is important to summarise our knowledge with respect to the temporal and spatial dimensions. An updated chronology of the pre-Hispanic metallurgy of Ecuador would look like this:

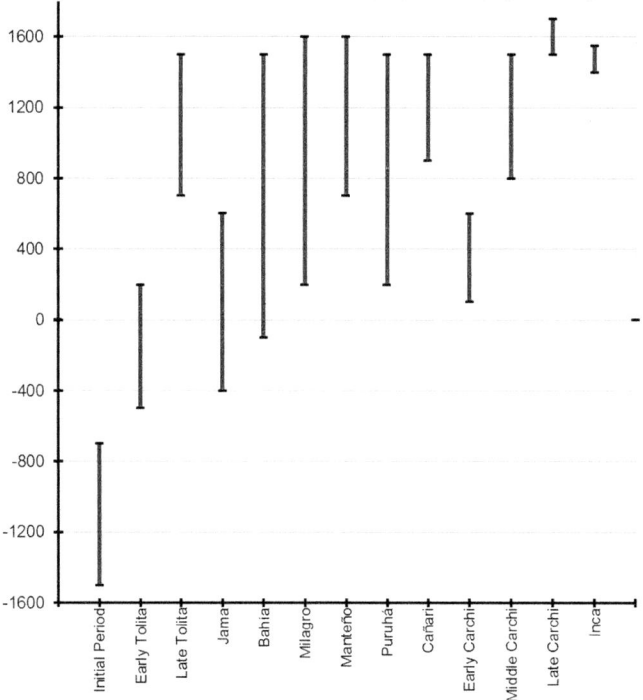

This global vision allows us to sketch a different way of looking at the development of metallurgy in Ecuador. To start, we need to acknowledge that in a wide area comprised between northern Chile and northern Ecuador, both in the coast and Sierra; there occurred during a long time span, probably between 2200 B.C. to 700 B.C. multiple social and technological processes that resulted in the invention of metallurgy in America. More than an original focus what seemed to exist was an invention area within which many foci experimented simultaneously with various metals and techniques.

Up to date, at least sixteen of these foci have been discovered. One of them is in northern Chile; Tulan 54, in the northern plateau (1200 B.C.) ((Graffam et al 1996) in González 2004.) In Bolivia there are the Wankarani, Chiripa and Pucara group in the Titicaca basin (1200 to 700 B.C.) and the early stages of Tiahuanaco (1000 B.C.) (González 2004). In Peru there are nine sites, including a recent discovery that turned out to be very early, Jiskairumoko (2155 to 1936 B.C.) (Aldenderfer et al 2008). The other Peruvian sites are: Waywaka in the Andahuaylas Sierra (1500 to 1000 B.C.); Mina Perdida in the Lurin Valley (1410 to 1090 B.C.); Puempaue in the Jequetepeque Valley (1500 to 1300 B.C.); Tanque in the coast (1000 B.C.); Kotosh in Cajamarca (1200 to 870 B.C.); Kuntur Wasi also in Cajamarca (900 B.C.); Cupisnique in the coast (1500 to 1200 B.C.) and Malpaso in the Lurin Valley (700 B.C.) (González 2004 and Lleras 2005).

The four remaining sites are in Ecuador: Putushio in Loja (1470 to 865 B.C.); Salango in Manabi (1500 B.C.); Los Cerritos in Santa Elena (890 B.C.) and Las Balsas in Esmeraldas (915 to 780 B.C.) (Lleras 2005). From our point of view there is no sense in starting a controversy about which of these sites could be the oldest and hence the "original focus". A we have said it seems more productive to think in terms of an area of invention where it could be possible to keep on finding sites of similar age. What is absolutely clear is that the territory of Ecuador shared in this ancient socio-technological dynamics playing as key role and not precisely as a passive receptor of inventions.

From this point onwards our proposal states that the developments attained during the Initial Period gave way to new socio-technological dynamics in the Ecuadorian coast and Sierra. In the north coast the technological, iconographic and formal patterns that formed the basis of the La Tolita – Tumaco Group were consolidated; the span of time between 500 B.C. and A.D. 200 is marked by an extraordinary production, the handling of platinum, gold and its alloys, copper and even lead. The La Tolita metallurgy irradiated its influence along the Pacific coast up to the Choco region, the Western Cordillera and the inter-Andean Cauca Valley; towards the south it influences the whole Manabi coast and the Guayas basin. Due in part to this expansive wave emerge the Calima – Ilama, Calima – Yotóco/Malagana and Early San Agustín/Tierradentro in Colombia as well as Jama – Coaque and Bahia in Ecuador.

While in the La Tolita area this influence gradually exhausts, in the Bahia area converge new trends that give this Group the vitality that Jama – Coaque never attained. Bahia metallurgy apparently brought together the influences of La Tolita with those of the initial focus of Santa Elena; the technological re-elaboration that happened there formed the basis for the late metallurgical nucleus of the coast. By the first centuries A.D. the basis of what would become the Milagro and Manteño metallurgies already existed. These two Groups, closely linked technologically and iconographically, re-took those previous experiences and started an expansive development that marked the history of metallurgy in the coast during much of the Common Era. The Milagro – Manteño influence reached the north coast where most of the characteristics of the Late Period of La Tolita are derived from it.

In the meanwhile the initial focus of the southern Sierra irradiated its influence over northern Peru and the central Ecuadorian Sierra; in the first centuries A.D. it is possible to see there most of the elements that form the Central Andean techno-formal complex (tupos, tumis, projectile points, head breakers, silver-copper alloy, gilded and silvered copper) but we cannot yet evaluate the role that the initial focus of Putushio played, though it is not likely that this huge development was just due to the Peruvian foci of Cajamarca. What is certain is that the Central Andean techno-formal complex appears in Cañari and Puruha in the first centuries A.D. so it cannot be the result of Moche or Vicus influence.

Cañari and Puruha are two Groups closely related, may be even closer than Milagro and Manteño. The extraordinary dynamics of these metallurgies in the southern and central Sierra defined the subsequent development of this industry, even in the north Sierra. Cañari-Puruha influence must have reached the Quito area rapidly, by the 3rd century there was in there a Minor Group that evolved locally and was the able to irradiate its influence northwards; first in Imbabura and then in Carchi and Nariño emerges the Early style that will develop into the Middle and Late styles that in Colombia appear as isolated, with no technological and iconographic relations to the other metallurgies of the country.

As a result of this complex and long dynamics, by the 15th century A.D. there were at least three Great Regional Groups in the coast (Milagro, Manteño and Late La Tolita) together with the residues of an older tradition (Bahia). In the Sierra there were the Puruha, Cañari and Carchi-Nariño Groups. Elsewhere, both in the coast and Sierra away from the predominance areas of the Great Groups there was metallurgical activity. The Inca invasion absorbed and integrated these multiple manifestations, not without taking from them several elements that gave rise to local hybrid styles.

This interpretative proposal can only account for the most general processes and is not detailed enough to explain many aspects and factors that would have to be examined within it in order to validate its explanatory efficiency. At least the bases have been set to conceive Ecuadorian metallurgy under a new light.

References

Aldenderfer, Mark, Nathan M. Craig, Robert J. Speakman and Rachel Popelka-Filcoff. 2008. *Four-thousand-year-old gold artifacts from the Lake Titicaca Basin*, southern Peru. PNAS, vol. 5, no. 13, April.

Alcina F. José.1975. *Excavaciones arqueológicas en Ingapirca (Ecuador)*. Mundo Hispánico, no. 38, Madrid.

Alcina F. José. 1979. *La arqueología de Esmeraldas, Ecuador*. Ministerio de Asuntos Exteriores, Madrid.

Bamps, Anatole. 1879. *Les antiquites equatoriennes du Musee Royal d'Antiquites de Bruxelles*. 3 Congres International des Americanistes, Bruxelles.

Barrandon, Jean Noel, Francisco Valdez y Patricia Estévez. 2002. *Identificación mineralógica de las fuentes del oro precolombino utilizado en la metalurgia prehispánica del Ecuador*. Sitoa, Madrid,

Bastian, Adolf.1878/86. *Die culturländer des alten America*, I-III. Berlin.

Benzoni, Girolamo. 1565/1985. *La historia del mondo nuovo*. Banco Central del Ecuador, Guayaquil.

Bergsøe, Paul.1937-38/1982. *Metalurgia y tecnología de oro y platino entre los indios precolombinos*. Editores Clemencia Plazas y Svend Bergsøe. Fondo de Copenhague y Cía. Metalúrgica Bera de Colombia S.A. Cali.

Bergsøe, Paul. 1938. *The gilding process and the metallurgy of copper and lead among the pre-Columbian Indians*. Ingeniorvidenskabelige Skrifter, no. a46, Copenhagen,

Bouchard, Jean François.1979. *Hilos de oro martillado hallados en la costa Pacífica del sur de Colombia*. Boletín Museo del Oro, año 2. Banco de la República. Bogotá.

Bouchard, Jean François. 1986. *Las más antiguas culturas precolombinas del Pacífico ecuatorial septentrional*. Miscelánea antropológica Ecuatoriana, no. 6, Banco Central del Ecuador, Guayaquil.

Bouchard, Jean François. 1995. *Arqueología de la costa del Pacífico nor-ecuatorial. Evaluación preliminar de los cambios ocurridos en los últimos decenios*. En: Primer encuentro de investigadores de la costa ecuatoriana en Europa. Editores: Aurelio Alvarez et al. Ediciones Abya-Yala.

Bray, Warwick.1971. *Los antiguos artífices americanos*. En: Conferencia Curl. Instituto de Arqueología. Londres.

Bray, Warwick. 2000. *Metal artefacts in the American world: archaeological evidence*. Manuscrito, Museo del Oro, Bogotá.

Bruhns, Karen. 1998. *Huaquería, procedencia y fantasía: los soles de oro del Ecuador*. Boletín Museo del Oro no. 44-45. Banco de la República, Bogotá.

Bushnell, G.H.S. 1951. The archaeology of the Santa Elena peninsula in south-west Ecuador. Cambridge.

Bustamante, Nohora, Lisette Garzón, Armando Bernal y Carlos Hernández. 2007. *Tecnología del platino en la fabricación de piezas de orfebrería precolombina*. Boletín del Museo del Oro no. 54, Bogotá.

Casas, Pablo.1991. *La Gorgona en tiempos precolombinos*. Revista de Antropología y Arqueología, vol. VII, nos. 1-2, Universidad de los Andes, Bogotá.

Cisneros, Fanny et al. 1998. *Museo del Banco Central del Ecuador, Ibarra.* Catalogo, Banco Central del Ecuador, Quito.

Christensen, Ross. 1954. *A recent excavation in southern coastal Ecuador.* Bulletin of the University Archaeological Society, 5, Brigham Young University, Utah.

Collier, Donald y John Murra. 1982. *Reconocimiento y excavaciones en el sur andino del Ecuador.* Centro de estudios históricos y geográficos del Azuay. Universidad Católica del Ecuador, sede en Cuenca, Cuenca.

Cortes, Emilia. 1997. *Tecnología y conservación de un ornamento prehispánico para la cabeza procedente de Nariño, Colombia.* Boletín Museo del Oro no. 43. Banco de la República. Bogotá.

Departamento Técnico Industrial (DTI). 2006. *Análisis de piezas Banco Central del Ecuador.* Ficha técnica dti-240-a, Banco de la República, Bogotá.

Di Capua, Constanza. 1997. *Una atribución cultural controvertida.* Fronteras de Investigación, año 1, no. 1, Quito.

Dorsey, George. 1901. *Archaeological investigations on the Island of La Plata, Ecuador.* Chicago.

Doyon, Leon. 2002. *Apuntes hacia un nuevo entendimiento de la historia cultural del área Carchi-Nariño.* Yale University, manuscript.

Duran, Miguel.1930. *Entierros en Huapán.* Revista de la Sociedad de Historia de Cuenca. no. 16, Cuenca.

Escalera, Andrés y María Angeles Barruiso.1978. *Estudio científico de los objetos de metal de Ingapirca (Ecuador).* Revista Española de Antropología Americana. 8.

Estévez de Romero, Patricia. 1994. *Estudio de objetos metálicos del sitio arqueológico La Tolita.* Manuscrito, Banco Central del Ecuador, Quito.

Estévez de Romero, Patricia. 1995. *Oro y platino en la orfebrería prehispánica del Ecuador.* Manuscrito, Banco Central del Ecuador, Quito.

Estévez de Romero, Patricia. 1998. *Platino en el Ecuador precolombino.* Boletín Museo del Oro no. 44-45. Banco de la República, Bogotá.

Estévez de Romero, Patricia. 1999. *Museo de esmeraldas. Informe de análisis de objetos de oro.* Manuscrito, Banco Central del Ecuador, Quito.

Estévez de Romero, Patricia. 2002. *Estudio analítico de la máscara funeraria: sol, cefalo, antropo, motivo zoo.* Manuscrito, Banco Central del Ecuador, Quito.

Estévez de Romero, Patricia. 2003. *Legado tecnológico en orfebrería.* Manuscrito, 51 Congreso Internacional de Americanistas, Santiago de Chile.

Estévez, Patricia, Antonio Fresco, Francisco Valdez y Alexandra Yépez. 2002. *Proyecto de metalurgia prehispánica.* Manuscrito, Banco Central del Ecuador, Quito.

Estrada, Emilio. 1957. *Ultimas civilizaciones prehistóricas de la cuenca del Río Guayas.* Museo Víctor Emilio Estrada, Guayaquil.

Estrada, Emilio. 1957. *Prehistoria de Manabí.* Publicaciones del Archivo Histórico del Guayas, Guayaquil.

Evans, Clifford and Betty J. Meggers. 1954. *Preliminary report on archaeological investigations in the Guayas basin, Ecuador.* Cuadernos de Historia y Arqueología, no. 12. Ecuador.

Farabee, William C. 1921. *A golden hoard from Ecuador.* The Museum Journal, University of Pennsylvania, March, vol. XII (1).

Fresco, Antonio.1984. *La arqueología de Ingapirca (Ecuador).* Quito.

García, Alfredo, Carolina Jervis y Pablo López. 2000. *El dorado de las narigueras en el Ecuador precolombino.* Miscelánea Antropológica Ecuatoriana, Revista del Museo Antropológico del Banco Central del Ecuador, año 9, no. 9, Guayaquil.

Garzón, Lissette, Armando Bernal y Carlos Hernández. 2007. *Nariño, algunos desarrollos tecnológicos en su orfebrería.* Metalurgia en la América Antigua, Roberto Lleras, editor, FIAN – IFEA, Bogotá – Lima.

Gómez del Corral, Luz Alba. 2007. *Desarrollo y simbolismo dual de la metalurgia de Nariño y Carchi.* Metalurgia en la América Antigua, Roberto Lleras, editor, FIAN – IFEA, Bogotá – Lima.

Gómez del Corral, Luz Alba y Roberto Lleras.1999. *La problemática de la arqueología nariñense vista desde la metalurgia.* Manuscrito, Simposio sobre metalurgia prehispánica, nuevas tendencias y perspectivas del Primer Congreso de Arqueología en Colombia, Manizales.

Gómez del Corral, Luz Alba y Roberto Lleras. 2006. *La problemática del Capuli, Piartal, Tuza: una nueva clasificación orfebre.* Humanidades, Revista de la Facultad de Ciencias Sociales y Humanas, año 2004, vol. 8, número 11-12, Universidad del Cauca, Popayán,

González, Luís R. 2004. *Bronces sin nombre. La metalurgia prehispánica en el noroeste argentino.* Ediciones Fundación Ceppa, Buenos Aires.

González Suarez, Federico.1904. *Prehistoria ecuatoriana.* Quito.

Guinea, Mercedes.1998. *La metalurgia del cobre en la costa norte del Ecuador durante el período de integración.* En: El área septentrional andina. Arqueología y etnohistoria. Compiladores: Mercedes Guinea, Jorge Marcos y J. F. Bouchard. Ediciones Abya-Yala. IFEA, Quito.

Heuzey, l. 1870. *Le tresor de Cuenca.* Gazette des beaux-arts, 4, Paris.

Holm, Olaf. 1963. *Copper needles from Manabí, Ecuador.* Ethnos 28 (2-4)

Holm, Olaf. 1966/7. *Money axes from Ecuador*, Folk, 8-9.

Holm, Olaf. 1968. *Fuelles que son unos cañutos (un comentario etno-arqueológico).* Editorial de la Casa de la Cultura Ecuatoriana, núcleo del Guayas. Guayaquil.

Holm, Olaf. 1970. *Orfebrería precolumbina del Ecuador.* Quito.

Holm, Olaf. 1977. *Lanzas silbadoras.* Pontificia Universidad Católica del Ecuador, Quito,

Hosler, Dorothy. 1998. *Los orígenes andinos de la metalurgia del occidente de México.* Boletín del Museo del Oro, 42.

Hosler, Dorothy, Heather Lechtman and Olaf Holm. 1990. *Axe-monies and their relatives.* Dumbarton Oaks, Washington. D.C.

Instituto Geográfico Militar del Ecuador. 2005. *Atlas multimedia del Ecuador.* IGME, Quito.

Instituto Geográfico Militar del Ecuador. Mapa físico-político del Ecuador. www.igme.gov.ec/downloads

Jijón y Caamaño, Jacinto. 1920. *Las tinculpas y notas acerca de la metalurgia de los aborígenes del Ecuador.* Boletín de la Academia Nacional de Historia, vol. 1. no. 1. Sociedad Ecuatoriana de Estudios Históricos. Quito.

Jijón y Caamaño, Jacinto. 1922. *La edad de bronce en América del Sur.* Boletín de la Academia Nacional de Historia, 4, Quito.

Jijón y Caamaño, Jacinto. 1997. *Antropología prehispánica del Ecuador*. Museo Jacinto Jijón y Caamaño. Embajada de España, Agencia de Cooperación Internacional. Editorial Santillana. Ediciones Abya-Yala. Quito.

Konanz, Max. 1944. *El arte entre los aborígenes de la provincia de Manabí*. Artes gráficas, Senafalder, Quito.

Lechtman, Heather.1986. *Traditions and styles in central Andean metalworking*. En: The beginning of the use of metals and alloys, Robert Maddin, editor, Cambridge.

Ledergerber, Paulina. 2004. *Ecuador: uno con el sol y la luna*. En: Simbolismo y ritual en los andes septentrionales. Editor: Mercedes Guinea. Editorial Complutense. Ediciones Abya-Yala. Quito.

Liu, Robert K. 1992. *Precolumbian personal adornment. La Tolita / Tumaco and Jama Coaque*. In: Ornament 16.

Lleras, Roberto. 2005. *Metales preciosos. Oro y plata de nuestros ancestros*. En: Joyas de los andes. Metales para los hombres, metales para los dioses. Museo Chileno de Arte Precolombino, Santiago de Chile.

Lleras, Roberto and Luz Alba Gómez. 1999. *The influence of central Andean metallurgy in the highlands of southern Colombia*. Manuscript, Symposium Metallurgy and Archaeology, IV World Archaeological Congress, Cape Town.

Lleras, Roberto, Luz Alba Gómez y Javier Gutiérrez. 2007. *El tiempo en los andes del norte de Ecuador y sur de Colombia: un análisis de la cronología a la luz de nuevos datos. Boletín del Museo Chileno de Arte Precolombino*, vol. 12, no. 1, Santiago de Chile.

Marcos, Jorge.1995. El mullo y el pututo. *La articulación de la ideología y el tráfico a larga distancia en la formación del estado Huancavilca*. En: Primer Encuentro de Investigadores de la Costa Ecuatoriana en Europa. Editores: Aurelio Alvarez et al. Ediciones Abya-Yala, Quito.

Marcos, Jorge. 1998. *A reassessment of the chronology of the Ecuadorian formative*. En: El área septentrional andina. Arqueología y etnohistoria. Compiladores: Mercedes Guinea, Jorge Marcos y J.F. Bouchard. Ediciones Abya-Yala. IFEA. Quito.

Maryon, Herbert. 1941. *Archaeology and metallurgy: II. The metallurgy of gold and platinum in pre-Columbian Ecuador*. Man, vol. 41, London.

Mayer, Eugen F. 1992. *Armas y herramientas de metal prehispánicas en Ecuador*. Philipp von Zabern, Mainz.

McCall, J.L. and R.D. Buchheit. 1971. *Metallographic studies of archaeological artifacts from Ecuador*. In Metallography. British Museum. London.

Meeks, Nigel, Susan La Niece and Patricia Estevez. 2002. *The technology of early platinum plating: a gold mask of the La Tolita culture*. Archaeometry, no. 44, part 2. University of Oxford, Oxford.

Meggers, Betty. 1966. *Ecuador. Ancient peoples and places*. vol. 49, Praeger Publishers, New York.

Mendoza Vargas, Sandra. 2002. *Las gentes y el oro en la costa Pacífica sur*. Manuscrito, Museo del Oro, Bogotá.

Meyers, Albert.1998. *Los incas en el Ecuador. Análisis de los restos materiales, I y II*. Colección Pendoneros, Instituto Otavaleño de Antropología, Banco Central del Ecuador, Abya-Yala, Quito.

Noguez, María, Rachel García, Guillermo Salas, Teresa Robert and José Ramírez. 2007. *About the pre-Hispanic Au-Pt "sintering" technique for making alloys*. International Journal of Powder Metallurgy, vol. 43, issue 1, Warrendale.

Ontaneda, Santiago. 2004. *Guión marco para la elaboración de los guiones museológicos de las salas de exhibición del nuevo Museo Nacional del Banco Central del Ecuador*. Manuscrito, Banco Central, Quito.

Ontaneda, Santiago y Antonio Fresco. 2002. *Museo del Banco Central del Ecuador, Riobamba*. Catalogo, Banco Central del Ecuador, Quito.

Oyarzun, Jorge. 2000. *Andean metallogenesis: a synoptical review and interpretation*. En Cordani et al, Tectonic evolution of South America, Rio de Janeiro.

Patiño Castaño, Diógenes. 1998. *Orfebrería prehispánica en la costa Pacífica de Colombia y Ecuador: "Tumaco-La Tolita"*, Boletín Museo del Oro no. 22. Banco de la República, Bogotá.

Patiño Castaño, Diógenes. 1993. *Más evidencias sobre orfebrería temprana en Tumaco-Tolita, costa Pacífica*. Boletín Museo del Oro no. 34-35. Banco de la República, Bogotá.

Patiño Castaño, Diógenes. 1997. *Arqueología y metalurgia en la costa Pacífica de Colombia y Ecuador*. Boletín Museo del Oro no. 43. Banco de la República. Bogotá.

Pérez, Juan Fernando et al. 1995. *Sala del Oro. Museo Nacional del Banco Central del Ecuador*. Catalogo, Quito.

Plazas, Clemencia. 1977/78. *Orfebrería prehistórica del altiplano nariñense, Colombia*. Revista Colombiana de Antropología, vol. XXI, ICAN, Bogotá.

Radiocarbon Database for Andes. n.d. *Geographical Index – Ecuador* www.archaeo.uw.edu.pl

Rehren, Thilo and Mathilde Themme. 1994. *Pre-Columbian gold processing at Putushio, south Ecuador: the archaeometallurgical evidence*. En Archaeometry of pre-Columbian sites and artifacts, David Scott, P. Meyers, editors, Getty Conservation Institute, Los Angeles

Reichlen, Henry. 1942. *Contribution a l'etude de la métallurgie précolombienne de la province d'Esmeraldas (Equateur)*. Journal de la Societe des Americanistes de Paris, vol. 34.

Reindel, Markus y Nicolas Guillaume-Gentil. 1995. *El proyecto arqueológico La Cadena. Estudios sobre la secuencia cultural de la cuenca del Río Guayas*. En: Primer Encuentro de Investigadores de la Costa Ecuatoriana en Europa. Editores: Aurelio Alvarez, Silvia G. Alvarez, Carmen Jauría y Jorge G. Marcos. Ediciones Abya-Yala, Quito.

Rivet, Paul. 1946. *Metalurgia del platino en la América precolombina*. En: Revista del Instituto Etnológico Nacional, Ministerio de Educación Nacional, Bogotá.

Rivet, Paul et Henry Arsandaux. 1946. *La metallurgie en Amerique précolombienne*. Travaux et memoirs de l'Institut de Etnologie, XXXIX, Paris.

Rodicio Garcia, Sara y Angel Riesco Terrero. 1998. *Minas de oro "Santa Bárbara" en los Cañaris*. En: El área septentrional andina. Arqueología y etnohistoria. Compiladores: Mercedes Guinea, Jorge Marcos y J.F. Bouchard. Ediciones Abya-Yala. IFEA. Ecuador.

Rovira, Salvador. 1990. *La metalurgia americana: análisis tecnológico de materiales prehispánicos y coloniales.* Editora de la Universidad Complutense de Madrid. Madrid.

Rovira, Salvador. 2004. *Un fragmento de placa dorada precolombina procedente de Ecuador: estudio analítico.* Anejos de AESPA, XXXII, Tecnología del oro antiguo: Europa y América. Concejo Superior de Investigaciones Científicas, Madrid.

Salgado, Héctor, David Stemper y Rolando Flórez. 1995. *Sociedades complejas en el litoral Pacífico: fragmentos de historia reconsiderados desde La Bocana.* En: Perspectivas regionales en la arqueología del suroccidente de Colombia y norte del Ecuador. Cristóbal Gnecco, editor. Universidad del Cauca, Popayán.

Saville, Marshall H. 1907 – 1910. *The antiquities of Manabí, Ecuador. Contributions to South American archaeology.* The George G. Heye Expedition. New York.

Saville, Marshall H. 1924. *The gold treasure of Sigsig, Ecuador.* Leaflets of the Museum of the American Indian. no. 3, New York.

Scott, David. 1982. *Pre-Hispanic Colombian metallurgy: studies of some gold and platinum alloys.* University of London, Institute of Archaeology. Ph.d. thesis, manuscript.

Scott, David. 1983. *Depletion gilding and surface treatment of gold alloys from the Nariño area of ancient Colombia.* Journal of the Historical Metallurgy Society, 17.

Scott, David. 1985. *Dorado por fusión y dorado de lámina en Colombia y Ecuador prehispánicos.* En: Metalurgia de América Precolombina. 45 Congreso de Americanistas. Colección Bibliográfica del Banco de la República. Bogotá.

Scott, David and Warwick Bray. 1980. *Ancient platinum technology in South America.* Platinum Metals Review, 24.

Scott, David and Warwick Bray. 1994. *Pre-Hispanic platinum alloys: their composition and use in Ecuador and Colombia.* In: Archaeometry of pre-Columbian sites and artefacts. Editors: David A. Scott and Peter Meyers. The Getty Foundation, Los Angeles,

Scott, David y Jean Francois Bouchard. 1988. O*rfebrería prehispánica de las llanuras del pacífico de Ecuador y Colombia.* Boletín Museo del Oro, no. 22. Banco de la República. Bogotá.

Scott, David y Jean Francois Bouchard. n.d. *Pre-Hispanic platinum alloys: their composition and utilisation in Ecuador and Colombia.* En: Tecnología en el mundo andino. Editors: Heather Lechtman and Ana María Soldi. México. Universidad Nacional Autónoma de México.

Scott, David y E. Doehne.1990. L*a soldadura con aleaciones de oro en la América antigua: un análisis de dos pequeños adornos provenientes del Ecuador.* Boletín Museo del Oro, no. 29. Banco de la República. Bogotá.

Segarra, G. 1924. *Comentarios y acotaciones al tesoro de Sigsig, Ecuador.* The gold treasure of Sigsig, Ecuador. Leaflets of the Museum of the American Indian no. 3, New York.

Stemper, David. 1993. *The persistence of pre-Hispanic chiefdoms on the Río Daule, coastal Ecuador.* University of Pittsburgh memoirs in Latin American archaeology no. 7, Pittsburgh, Quito.

Stemper, David y Héctor Salgado. 1993. *Metalurgia prehispánica y colonial-republicana en el Pacífico colombiano.* Revista Colombiana de Antropología, vol. XXX, ICAN, Bogotá.

Stothert, Karen E. 1997. *Fundición tradicional campesina en la costa del Ecuador.* Boletín Museo del Oro no. 43. Bogotá.

Subdirección Técnica, 2005. *Arqueometalurgia.* Base de datos, Museo del Oro. Bogotá

Sutliff, Marie J. s.f. *El proceso productivo metalúrgico de la cultura Milagro: el caso de Peñón del Río.* Tesis de Licenciatura, Escuela Superior Politécnica del Litoral. Guayaquil.

Sutliff, Marie J. 1989. *Domestic production of small copper artefacts during the Milagro occupation at Peñón del Rio (Guayas basin).* Centro de Estudios Arqueológicos y Antropológicos ESPOL, Guayaquil.

Sutliff, Marie J. 1990. *Contextual evidence for non – elite metallurgical production and use in Milagro society.* Centro de Estudios Arqueológicos y Antropológicos ESPOL, Guayaquil.

Ubelaker, Douglas. 1981. *The Ayalan cemetery: a late intermediate period burial site on the south coast of Ecuador.* Smithsonian Contributions to Anthropology, no. 29, Smithsonian Institution Press, Washington.

Uhle, Max. 1922. *Sepulturas ricas de oro en la provincia del Azuay.* Boletín de la Academia Nacional de Historia, tomo IV, no. 9, Quito.

Unknown author. n.d. *Una agrupación tecnológica de hachas metálicas del Ecuador en función de criterios morfológicos.* Manuscrito, Museo del Oro, Bogotá.

Valdez, Francisco. 1987. *Proyecto arqueológico La Tolita.* Catalogo exposición, Banco Central del Ecuador, Quito.

Valdez, Francisco.1989. *Le symbolisme du naturel et du social.* En Equateur, la terre et l'or, Maison de l'Amerique Latine, Paris.

Valdez, Francisco y Diego Veintimilla. 1992. *Signos amerindios. 5.000 años de arte precolombino en el Ecuador.* Dinediciones, Quito.

Valdez, Francisco, Patricia Estévez y Jean Noel Barrandon. 2007. *Mucho ruido y pocas nueces. El epílogo de la controversia del origen de los soles de oro del Ecuador.* En: Metalurgia en la América Antigua, Roberto Lleras, editor, FIAN – IFEA, Bogotá – Lima.

Valdez, Francisco, Bernard Gratuze, Alexandra Yépez y Julio Hurtado. 2007. *Evidencia temprana de metalurgia en la costa Pacífica ecuatorial.* Boletín del Museo del Oro, no. 53, Bogotá.

Verneau, Renée et Paul Rivet.1912. *Ethnographie ancienne de l'Equateur.* Gauthier-Villars, Paris.

Von Buchwald, Otto. 1918. *Notas acerca de la arqueología del Guayas.* Boletín de la Sociedad Ecuatoriana de Estudios Históricos Americanos, no. 1, Ecuador.

Wolf, Teodoro. 1879. *Viajes científicos por la Republica del Ecuador, III. Memorias sobre geografía y geología de la provincia de Esmeraldas.* Imprenta del Comercio, Guayaquil.

Yépez, Alexandra. 2000. *Museo Regional de Esmeraldas.* Catálogo, Banco Central del Ecuador, Quito.

Zapater, Irving Iván. s.f. *Mecenas de la cultura.* Manuscrito, Banco Central del Ecuador, Quito.

Zevallos, Carlos. 1956. *Tecnología metalúrgica arqueológica. La elaboración del alambre.* Cuadernos de Historia y Arqueología, no. 16-18, Quito.

Zevallos, Carlos. 1965/66. *Estudio regional de la orfebrería precolombina de Ecuador y su posible relación con las áreas vecinas.* Revista del Museo Nacional. Tomo XXXIV. Quito.